- PYTHON -

5 Manuscripts in 1 Book

Learn Coding Programs with Python Programming
and Master Data Analysis & Analytics, Data Science
and Machine Learning with the Complete Crash
Course for Beginners

by

TechExp Academy

2020 © Copyright

Manuscript 1: Python For Beginners

Manuscript 2: Advanced Python Programming

Manuscript 2: Python for Data Analysis and Analytics

Manuscript 3: Python for Data Science

Manuscript 4: Python Machine Learning

Consolidated Table of Contents

Copyright, Legal Notice and Disclaimer

- PYTHON FOR BEGINNERS -

The Python Programming Crash Course for Beginners to Learn Python Coding Well in 1 Week with Hands-On Exercises

by

TechExp Academy

2020 © Copyright

Table of Contents

Chapter 1: Intro to Python

This book is about Python for beginners. It introduces the core aspects of the Python programming language. Python is a high-level, integrated, general-purpose programming language developed in 1991 by Guido van Rossum. The design philosophy underpinning Python emphasizes on code readability characterized by its use of considerable whitespace. Python's object-oriented approach and language constructs focus on helping programmers write logical, clear code for large and small-scale projects. This book aims to provide a cogent introduction to python for beginners. It seeks to provide a platform to learn python programming well and in one week including step by step practical examples, exercises, and tricks.

Before we go into the' why programming' discussion first let us knows what programming is. Programming is the process of taking an equation and converting it into a code, a programming language so that it is implemented on a machine. Or in simple words, "Programming is a language to teach a machine what to do through a set of instructions." Various types of programming languages are used, for example:

- Python

- PHP

- C language

- JAVA and more.

So why is it important? What's so Great about this? Why does it matter?

The first thing, everything is done on computers in today's world; from sending an email / report to a distant colleague / friend / relative or even as simple as a picture to having an important meeting on Skype! It has become a necessity for every single individual to have a computer / laptop as it is fast, very reliable and easy to use. So when computers are part of your life, it's said that learning to program will boost your life! One of the main reason people are learning programming is because they want to make a career by creating websites for companies or mobile apps. That is not the only reason you need to learn programming; programming can also help to improve a person's efficiency and productivity!

What is Python?

Python is a multi-purpose language created by Guido van Rossum. Differently from HTML, CSS, and JavaScript, Python can be used for multiple types of programming and software development. Python can be used for things like:

- Back end software development
- Desktop apps
- Big data and mathematical computations
- System scripts
- Data Analysis
- Data Science
- Artificial Intelligence
- And many others

The reasons why Python is the go-to language for many can be summarized in the following points:

- Beginner Friendliness as it reads like almost English
- Very no hard rules on how to build features
- Easier to manager error
- Many and big supporting communities
- Careers opportunities
- Future especially for the strong fit towards Data Science and Machine Learning Applications.

Python is one of the most powerful programming/coding languages. It's one of the programming languages that are interpreted rather than compiled. This means the Python Interpreter works or operates on Python programs to give the user the results. The Python Interpreter works in a line-by-line manner. With Python, one can do a lot. Python has been used for the development of apps that span a wide of fields, from the most basic apps to the most complex ones. Python can be used for the development of the basic desktop computer applications. It is also a good coding language for web development. Websites developed with Python are known for the level of security and protection they provide, making them safe and secure from hackers and other malicious users. Python is well applicable in the field of game development. It has been used for the development of basic and complex computer games. Python is currently the best programming language for use in data science and machine learning. It has libraries that are best suitable for use in data analysis, making it suitable for use in this field. A good example of such a library is scikit-learn (sklearn) which has proved to be the best for use in data science and machine learning.

Python is well known for its easy-to-use syntax. It was written with the goal of making coding easy. This has made it easy the best language even for beginners. Its semantics are also easy, making it easy for one to understand Python codes. The language has received a lot of changes and improvements, especially after the introduction of Python 3. Previously, we had Python 2.7 which had gained much stability. Python 3 brought in new libraries, functions, and other features, and some of the language's construct were changed significantly.

Features of Python

Python is Easy

Python is an easy language to get started with. The ease of use is underscored by the fact that most of the programs written in python look similar to the English language. Therefore, this simplicity makes python an ideal learning language for entry-level programming courses, and thus introducing programming concepts to students.

Python is Portable/Platform Independent

Python is massively portable, which means that Python programs can be run in various operating systems without needed specific or extensive changes.

Python is an Interpreted Language

Mainly, Python is an interpreted language as opposed to being a compiled language, on the other hand, C, C++ are examples of complied languages.

In many cases, the programs composed in a high-level language are typically referred to as source code or source programs. As a result, the commands in the source codes are referred to as statements. A computer lacks the capacity to execute a program written in high-level language. Typically, computers understand machine language, which comprises of 1s and 0s only.

As a result, there are two types of programs available to users when translating high-level languages to machine languages: compiler and interpreter.

Compiler

In its operative function, a complier translates the entire source code into the readable machine language in one swoop. The machine language is subsequently executed. The process is illustrated below:

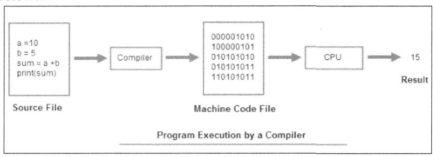

Program Execution by a Compiler

Interpreter

On the other hand, an interpreter employs a line by line approach to translating a high-level language into machine language, which is subsequently executed. The Python interpreter bags are at the top of the file, translates the beginning line into machine language, and subsequently executes it. The process repeatedly continues throughout the entire file as illustrated below:

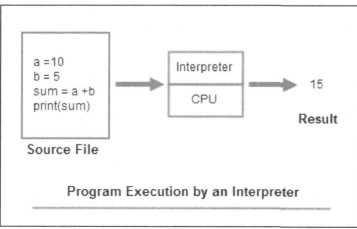

Program Execution by an Interpreter

It is crucial to distinguish between high-level and complied languages. For example, the compiled languages such as C, and C++ use a compiler to translate and interpret (high-level to machine language.) On the other hand, an interpreted language such as Python employs an interpreter to conduct this approach of translation and subsequent execution. The important distinction here between interpreted and compiled language is that the generally complied language operates and performs better compared to written programs that employ interpreted languages. However, Python does not suffer from this disadvantage.

In synthesis, Python is an interpreted language as the program executes directly from the source code. Every time Python programs are run the source code is required. What Python does is to convert the source code written by the developer into an intermediate code which is then further converted into the machine language ready to be executed. Python is therefore an Interpreted language as it is processed in real time by the interpreter. The source code does not need to be compiled before its execution. Compiled means that the code needs to be converted to machine language before runtime.

Python is Dynamically Typed

Another characteristic of Python is to be dynamically typed which means that data types are checked on the fly, during execution vs statically typed when data types are checked before run-time.

Python is Strongly Typed

The primary feature of a strongly typed language is that it lacks the capacity to convert one form (type) of data to another automatically. On the other hand, languages such as PHP and JavaScript, which are loosely typed, have the capacity to convert data from one type to another freely and automatically. Consider the following case:

```
1  price = 12
2  str = "The total price = " + 12
3  console.log(str)
```

Output:

```
The total price = 12
```

In this regard, before adding 12 to the string, the JavaScript language seeks to convert the number 12 to a string "12", which is subsequently appended to the end of the string. However, in a Python statement such as below:

```
str = "The total price = " + 12
```

The language (Python) would occasion an error because it does not have the capacity to convert the umber 12 into a string.

In synthesis, Python is a strongly typed language i.e. the type of variables must be known. This implies that the type of a value doesn't change in unexpected ways. For example, a string containing only a number doesn't automatically change into a number, as may happen in Perl. In Python, every data type change needs an explicit conversion.

A huge set of libraries

Python provides users with a broad range of libraries, which make it easy to add new capacities and capabilities without necessarily reinventing new approaches. The following are common questions mostly posed from beginners in Python programming:

What type of application I can create using Python?

Some of the main applications for which Python is used include:
Games
- Machine Learning and Artificial Intelligence
- Data Science and Data Visualization
- Web Development
- Game Development

- Desktop GUI
- Web Scraping Applications
- Business Applications
- Audio and Video Applications
- CAD Applications
- Embedded Applications
- System administration applications
- GUI applications.
- Console applications
- Scientific applications
- Android applications

And several others not mentioned here.

Who uses Python?

Some of the main companies that utilize Python include:

i. YouTube

ii. Mozilla

iii. Dropbox

iv. Quora

v. Disqus

vi. Reddit

vii. Google

viii. Disney
ix. Mozilla
x. Bit Torrent
xi. Intel
xii. Cisco
xiii. Banks such as JPMorgan Chase, UBS, Getco, and Citadel apply Python for financial market forecasting.
xiv. NASA for scientific programming tasks
xv. iRobot for commercial robotic vacuum cleaner
xvi. And many others.

Chapter 2: Installing Python

To code in Python, you must have the Python Interpreter installed on your computer. You must also have a text editor in which you will be writing and saving your Python codes. The good thing with Python is that it can run on various platforms like Windows, Linux, and Mac OS. Most of the current versions of these operating systems come installed with Python. You can check whether Python has been installed on your operating system by running this command on the terminal or operating system console:
Python

Type the above command on the terminal of your operating system then hit the Enter/Return key.

The command should return the version of Python installed on your system. If Python is not installed, you will be informed that the command is not recognized; hence you have to install Python.

Choosing a Python Version

The main two versions of Python are 2.x and 3.x. Python 3.x is obviously the latest one but Python 2.x as of today is most likely still the most used one. Python 3.x is however growing much faster in terms of adoption. Python 2.x is still in use in many software companies. More and more enterprises however are moving to Python 3.x. There are several technical differences between the 2 versions. We can summarize in very a simple way as Python 2.x is legacy and Python 3.x is the future. The advice for you is to go for the latest version Python 3.x. From 2020 Python 2.x is not be supported anymore.

General installations instruction

Installing Python is very easy. All you need to do is follow the steps described below:
1) Go to Python downloads page
https://www.python.org/downloads/
2) Click the link related to your operating system

3) Click on the latest release and download according to your operating system

4) Launch the package and follow the installation instructions (we recommend to leave the default settings) Make sure you click on Add Python 3.x to PATH. Once the installation is finished you are set to go!

5) Access your terminal IDLE

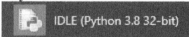

Test that all works by writing your first Python code:

> print ("I'm running my first Python code")

Press enter or return, this is what you should get

```
>>> print ("I'm running my first Python code")
I'm running my first Python code
```

You can do the same also by launching this command using a file. We will address this after we address the Python IDLE or another code editor.

Installation on Windows

To install Python on Windows, download Python from its official website then double click the downloaded setup package to launch the installation. You can download the package by clicking this link:

https://www.python.org/downloads/windows/

It will be good for you to download and install the latest package of Python as you will be able to enjoy using the latest Python packages. After downloading the package, double click on it and you will be guided through on-screen instructions on how to install Python on your Windows OS.

Installation on Linux (Ubuntu)

In Linux, there are a number of package managers that can be used for installation of Python in various Linux distributions. For example, if you are using Ubuntu Linux, run this command to install Python:

$ sudo apt-get install python3-minimal

Python will be installed on your system. However, most of the latest versions of various Linux distributions come installed with Python. Just run the "python" command. If you get a Python version as the return, then Python has been installed on your system. If not, go ahead and install Python.

Installation on Mac OS

To install Python in Mac OS, you must first download the package. You can find it by opening the following link on your web browser:

https://www.python.org/downloads/mac-osx/

After the setup has been downloaded, double click it to launch the installation. You will be presented with on-screen instructions that will guide you through the installation process. Lastly, you will have Python running on your Mac OS system.

Running Programs

One can run Python programs in two main ways:
- Interactive interpreter
- Script from the command line

Interactive Interpreter or Interactive Mode via Shell

Python comes with a command-line which is commonly referred to as the interactive interpreter. You can write your Python code directly on this interpreter and press the enter key. You will get instant results. If you are on Linux, you only have to open the Linux terminal then type "python". Hit the enter key and you will be presented with the Python interpreter with the >>> symbol. To access the interactive Python interpreter on Windows, click Start -> All programs, then identify "Python ..." from the list of programs. In my

case, I find "Python 3.5" as I have installed Python 3.5. Expand this option and click "Python ...". In my case, I click "Python 3.5(64-bit)" and I get the interactive Python interpreter.

Here, you can write and run your Python scripts directly. To write the "Hello" example, type the following on the interpreter terminal:
print("Hello")
Hit the enter/return key and the text "Hello" will be printed on the interpreter:

```
>>> print("Hello")
Hello
>>>
```

Script from Command Line

This method involves writing Python programs in a file, then invoking the Python interpreter to work on the file. Files with Python should be saved with a .py extension. This is a designation to signify that it is a Python file. For example, script.py, myscript.py, etc. After writing your code in the file and saving it with the name "mycode.py", you can open the operating system command line and invoke the Python interpreter to work on the file. For example, you can run this command on the command line to execute the code on the file mycode.py:

> python mycode.py

The Python interpreter will work on the file and print the results on the terminal.

Python IDE (Integrated Development Environment)

If you have a GUI (Graphical User Interface) application capable of supporting Python, you can run the Python on a GUI environment. The following are the Python IDEs for the various operating systems:

- UNIX- IDLE
- Windows- PythonWin

Macintosh comes along with IDLE IDE, downloadable from the official website as MacBinary or BinHex'd files.

Chapter 3: IDLE and Python Shell

As we have seen Python can be used in 2 main modes:
1) The interactive mode also known as via Python Shell
2) Via Python IDLE

As a reminder, the Python Shell known as the prompt string is ready to accept commands. The Python shell allows typing Python code and get the result immediately. The Python Shell is good for testing a small chunk of code. The Python IDLE (IDLE or IDE stands for Integrated Development Environment) instead allows to do that as well but gives much more functionalities. Therefore we advise you to go straight for the Python IDLE. In order to start using Python IDLE we recommend also creating a new directory for example "Python-Practice" where you wish, in there you will save future python files.

Installing the Interpreter

As a reminder, if not done yet before we can write our first Python program, we have to download the appropriate interpreter for our computers.
We'll be using Python 3 in this book because as stated on the official Python site "Python 2.x is legacy; Python 3.x is the present and future of the language". In addition, "Python 3 eliminates many quirks that can unnecessarily trip up beginning programmers". However, note that Python 2 is currently still rather widely used. Python 2 and 3 are about 90% similar. Hence if you learn Python 3, you will likely have no problems understanding codes written in Python 2.

To install the interpreter for Python 3, head over to
https://www.python.org/downloads/.

The correct version should be indicated at the top of the webpage. We'll be using version 3.6.1 in this book. Click on "Download Python 3.6.1" and the software will start downloading.

Alternatively, if you want to install a different version, scroll down the page and you'll see a listing of other versions. Click on the release version that you want. You'll be redirected to the download page for that version.

Scroll down towards the end of the page and you'll see a table listing various installers for that version. Choose the correct installer for your computer. The installer to use depends on two factors:
1. The operating system (Windows, Mac OS, or Linux) and
2. The processor (32-bit vs 64-bit) that you are using.

For instance, if you are using a 64-bit Windows computer, you will likely be using the "Windows x86-64 executable installer". Just click on the link to download it. If you download and run the wrong installer, no worries. You will get an error message and the interpreter will not install. Simply download the correct installer and you are good to go. Once you have successfully installed the interpreter, you are ready to start coding in Python.

Python IDLE

We advise using at least the default Python IDLE. However, there are many other options you can use the table below to compare them. IDLE is the integrated development environment for Python and it is installed automatically with Python. As well as a neat graphical user interface, IDLE is packed with features that make using Python for developing easy, including a very powerful feature, syntax highlighting.
With syntax highlighting, reserved keywords, literal text, comments, etc. are all highlighted in different colors, making it much easier to see errors in your code. As well as editing your Python program with IDLE, you can also execute your programs in IDLE.

Features	IDLE	Thonny	Eric Python	Atom	Wing	Sublime	Rodeo	PyDev	Spyder	PyCharm
Code Completion	✗	✓	✓	✓	✓	✓	✓	✓	✓	✓
Debugging	✓	✓	✓	Package Available	✓	Package Available	✗	Remote Debugger	✓	✓
Built-in Unit Testing	✗	✗	✓	Package Available	✓	Package Available	✗	✓	Plugin	✓
Open Source	✓	✓	✓	✓	✗	✗	✓	✓	✓	Community Edition
Light Weight	✓	✓	✗	✓	✓	✓	✗	✓	✗	✗
Refactoring	✗	✗	✗	Package Available	✓	Package Available	✗	✓	✓	✓

When you install Python, you get IDLE, an integrated IDE. To start it in Windows, find the ArcGIS folder on your computer. Inside the Python folder, you will see IDLE as a choice – to start IDLE, select it. IDLE has an interactive interpreter built-in and can easily run full scripts. The GUI module built into Python is used to write IDLE so it is the same language as the one it will execute.

IDLE has another advantage over python.exe in that script outputs, including print statements, are sent straight to the interactive window in IDLE, and this doesn't disappear once the script has been executed. IDLE also uses little memory.

There are disadvantages too; not much in the way of code assist, i.e. autocomplete, and not much in the way of organization in terms of logical projects. Every variable in a script cannot be located like in other IDEs and there is a limit to the number of scripts listed in the Recent Files menu – this is a bit of an obstruction when it comes to finding scripts that haven't been run for a while. IDLE isn't a bad IDE, it is ideal if there aren't any other options and it can be used to quickly test a snippet of code.

Using the Python Shell, IDLE and Writing our FIRST program

We'll be writing our code using the IDLE program that comes bundled with our Python interpreter. To do that, let's first launch the IDLE program. You launch the IDLE program like how you launch any other programs. For instance, on Windows 10, you can search

for it by typing "IDLE" in the search box. Once it is found, click on IDLE (Python GUI) to launch it. You'll be presented with the Python Shell shown below.

The Python Shell allows us to use Python in interactive mode. This means we can enter one command at a time. The Shell waits for a command from the user, executes it, and returns the result of the execution. After this, the Shell waits for the next command.

Try typing the following into the Shell. The lines starting with >>> are the commands you should type while the lines after the commands show the results. In this book we will indicate with this symbol ">" commands to be written and with this "→" the results.

>   2+3

→ 5

>   3>2

→ True

>   print ('Hello World')

→ Hello World

When you type 2+3, you are issuing a command to the Shell, asking it to evaluate the value of 2+3. Hence, the Shell returns the answer 5. When you type 3>2, you are asking the Shell if 3 is greater than 2. The Shell replies True. Next, print is a command asking the Shell to display the line Hello World.

Python Interactive Mode

Python Shell is a very convenient tool for testing Python commands, especially when we are first getting started with the language. However, if you exit from the Python Shell and enter it again, all the commands you type will be gone. Also, you cannot use the Python Shell to create an actual program. To code an actual program, you need to write your code in a text file and save it with a .py extension. This file is known as a Python script.

Python Script Mode

To create a Python script, click on File > New File in the top menu of our Python Shell. This will bring up the text editor that we are going to use to write our very first program, the "Hello World" program. Writing the "Hello World" program is kind of like the rite of passage for all new programmers. We'll be using this program to familiarize ourselves with the IDLE software.

Type the following code into the text editor (not the Shell).

```
>    #Prints the Words "Hello World"
>    print ("Hello World")
→    Hello World
```

You should notice that the line #Prints the Words "Hello World" is in red while the word print is in purple and "Hello World" is in green. This is the software's way of making our code easier to read. The words print and "Hello World" serve different purposes in our program, hence they are displayed using different colors. We'll go into more details in later chapters. The line #Prints the Words "Hello World" (in red) is actually not part of the program. It is a comment written to make our code more readable for other programmers. This line is ignored by the Python interpreter. To add comments to our program, we type a # sign in front of each line of comment, like this:

> #This is a comment

> #This is also a comment

> #This is yet another comment

Alternatively, we can also use three single quotes (or three double quotes) for multiline comments, like this:

> '''

This is a comment

This is also a comment

This is yet another comment

'''

Now click File > Save As... to save your code. Make sure you save it with the .py extension.
Done? Voilà! You have just successfully written your first Python program.
Finally click on Run > Run Module to execute the program (or press F5). You should see the words Hello World printed on your Python Shell.

How to Write a Python Program and Run it in IDLE

1. Start IDLE – open Start>All Programs>Python>IDLE

2. A window with a title of Python Shell will open

3. Click on File>New Window

4. Now a new window called Untitled will load

5. Click on File>Save As and choose a location for your program file

6. Where it says File Name, type program1/py in the box

7. Click on Save

8. A blank window will open – this is an editor window and it is ready for you to type your program in.

9. Type the following statement in exactly as written – it will work on Python 2.x or 3.x:

print ("Hello World")

10. Open the Run menu and click on Run Module to run the program

11. You will now see a message asking you to save your program (it will say Source) so click OK

12. Your program will now run in a Python Shell window

13. To quit Python, shut down all Python windows

Important Note

If you want to open your file again, find it in the folder you saved it in. Right-click on it and then choose Edit with IDLE from the menu – this will open the editor window

Other Python Interactive Developer Environment (IDE)

One of the best IDE to use with Python is Eclipse so, if you wish you can install it on your computer.

1. Go to http://www.eclipse.org/downloads and download the Eclipse installation. It will be in zip file format.

2. Unzip the file and that is it, nothing else needs to be done; Eclipse will be installed on your system.

3. To start Eclipse, go to the directory where you unzipped the file and double-click on eclipse.exe.

The Eclipse Python Plugin

PyDev is an Eclipse Python IDE and it can be used in several different Python distributions. It supports graphical debugging, code refactoring, code analysis, and lots more besides. PyDev can be installed through the Eclipse update manager by going to http://pydev.org/updates. Just check the box beside PyDev and follow the on-screen instructions to install it.

Next, you need to configure Eclipse because it needs to know where Python is.

1. Open Window>Preferences

2. Click on the option for PyDev and then click on Interpreter Python

3. Click New Configuration

4. Add the executable path for Python

Eclipse IDE is now set up on your computer and ready to use with Python.

Types of Errors

In programming are in Python 3 main types of errors can be encountered:

- Syntax Errors.
- Runtime Errors.
- Logical Errors.

Syntax Errors

The syntax is a set of guidelines that need to be followed to write correctly in a computer language. A syntax error is a missing punctuation character, a mistake such as a misspelled keyword, or a missing closing bracket. The syntax errors are detected by the compiler or the interpreter. If you try to execute a Python program that contains a syntax error, you will get an error message on your screen and the program won't execute. You must correct any errors and then try to execute the program again. Usually these types of errors are due to typos. In case an error occur the Python interpreter stops running. Some commons causes of syntax errors are due to:

- Wrongly written keywords
- Wrong use of an operator
- Forgetting parentheses in a function call
- Not putting strings in single quotes or double quotes

Runtime Errors

These errors occur when the execution of the code is stopped because of an operation being impossible to be carried out. A runtime error can cause a program to end abruptly or even cause a system shut-down. Such errors can be the most difficult errors to detect. Running out of memory or a division by zero are examples of runtime errors.

Logical Errors

Logical errors happen when the code produces wrong results. For instance a temperature conversion from Fahrenheit to Celsius

→ 1   → print("20 degree Fahrenheit in degree Celsius is: ")
→ 2   → print(5 / 9 * 20 - 32)

Result
→ 1   → 20 degree Fahrenheit in degree Celsius is:
→ 2   → -20.88888888888889

The above code outputs -20.88888888888889, that is erroneous. The right result is -6.666. These types of errors are referred as logical errors. To get the correct answer parenthesis needs to be used correctly 5 / 9 * (20 - 32) instead of 5 / 9 * 20 - 32.

A logic error is an error that prevents your program from doing what you expected it to do. With logic errors, you get no warning at all. Your code compiles and runs but the result is not the expected one. You must review your program thoroughly to find out where your error is. Python program executes as normal. It is the programmer who has to find and correct the erroneously written Python statement, not the computer or the interpreter. Computers are not that smart after all.

Chapter 4: Data Types and Variables in Python

Every program has certain data that allows it to function and operate in the way we want. The data can be a text, a number, or any other thing in between. Whether complex or as simple as you like, these data types are the cogs in a machine that allow the rest of the mechanism to connect and work. Python is a host to a few data types and, unlike its competitors; it does not deal with an extensive range of things.

That is good because we have less to worry about and yet achieve accurate results despite the lapse. Python was created to make our lives, as programmers, a lot easier.

Strings

In Python, and other programming languages, any text value that we may use, such as names, places, sentences, they are all referred to as strings. A string is a collection of characters, not words or letters, which is marked by the use of single or double quotation marks. To display a string, use the print command, open up a parenthesis, put in a quotation mark, and write anything.

Once done, we generally end the quotation marks and close the bracket. If you are using PyCharm, the IntelliSense detects what we are about to do and delivers the rest for us immediately. You may have noticed how it jumped to the rescue when you only type in the opening bracket. It will automatically provide you with a closing one. Similarly, for the quotation marks, one or two, it will provide the closing one's for you.

See why we are using PyCharm?

It greatly helps us out. "I do have a question. Why do we use either single or double quotation marks if both provide the same result?" Ah! Quite an eye.

There is a reason we use these, let me explain by using the example below:

print('I'm afraid I won't be able to make it')
print("He said "Why do you care?"")

Try and run this through PyCharm.

Remember, to run, simply click on the green play-like button on the top right side of the interface.

> "C:\Users\Programmer\AppData\Local\Programs\Python\Python37-32\python.exe"
> "C:/Users/Programmer/PycharmProjects/PFB/Test1.py"

> File "C:/Users/Programmer/PycharmProjects/PFB/Test1.py", line 1

> print('I'm afraid I won't be able to make it')

→ SyntaxError: invalid syntax

Process finished with exit code 1
Here's a hint: That's an error!
- So what happened here?
- Try and revisit the inputs.
- See how we started the first print statement with a single quote?
- Immediately, we ended the quote using another quotation mark.
- The program only accepted the letter 'I' as a string.

You may have noticed how the color may have changed for every other character from 'm' until 'won', after which the program detects yet another quotation mark and accepts the rest as another string. Quite confusing, to be honest.

Similarly, in the second statement, the same thing happened. The program saw double quotes and understood it as a string, right until the point the second instance of double quotation marks arrives. That's where it did not bother checking whether it is a sentence or that it may have still been going on. Computers do not understand English; they understand binary communications. The compiler is what runs when we press the run button. It compiles our code and interprets the same into a series of ones and zeros so that the computer may understand what we are asking it to do.

This is exactly why the second it spots the first quotation mark, it considers it as a start of a string, and ends it immediately when it spots a second quotation mark, even if the sentence was carrying onwards. To overcome this obstacle, we use a mixture of single and double quotes when we know we need to use one of these within the sentence. Try and replace the opening and closing quotation marks in the first state as double quotation marks on both ends. Likewise, change the quotation marks for the second statement to single quotation marks as shown here:

> print("I'm afraid I won't be able to make it")

> print('He said "Why do you care?"')

Now the output should look like this:
→ I'm afraid I won't be able to make it
→ He said, "Why do you care?"

Lastly, for strings, the naming convention does not apply to the text of the string itself. You can use regular English writing methods and conventions without worries, as long as that is within the quotation marks. Anything outside it will not be a string in the first place, and will or may not work if you change the cases.

Numeric Data type

Python is able to recognize numbers rather well. The numbers are divided into two pairs:
- Integer – A positive and/or negative whole numbers that are represented without any decimal points.
- Float – A real number that has a decimal point representation.

This means, if you were to use 100 and 100.00, one would be identified as an integer while the other will be deemed as a *float*. So why do we need to use two various number representations?

If you are designing a program, suppose a small game that has a character's life of 10, you might wish to keep the program in a way that whenever a said character takes a hit, his life reduces by one or two points. However, to make things a little more precise, you may need to use float numbers. Now, each hit might vary and may take 1.5, 2.1, or 1.8 points away from the life total. Using floats allows us to use greater precision, especially when calculations are on the cards. If you aren't too troubled about the accuracy, or your programming involves whole numbers only, stick to integers.

Booleans in Python

Ah! The one with the funny name. Boolean (or bool) is a data type that can only operate on and return two values: True or False. Booleans are a vital part of any program, except the ones where you may never need them, such as our first program. These are what allow programs to take various paths if the result is true or false.

Here's a little example. Suppose you are traveling to a country you have never been to. There are two choices you are most likely to face. If it is cold, you will be packing your winter clothes. If it is warm, you will be packing clothes that are appropriate for warm weather. Simple, right? That is exactly how the Booleans work. We will look into the coding aspect of it as well. For now, just remember, when it comes to true and false, you are dealing with a bool value.

List Python

While this is slightly more advanced for someone at this stage of learning, the list is a data type that does what it sounds like. It lists objects, values, or stores data within square brackets ([]). Here's what a list would look like:

```
>  month = ['Jan', 'Feb', 'March', 'And so on!']

>  month

→  ['Jan', 'Feb', 'March', 'And so on!']
```

We will be looking into this separately, where we will discuss lists, tuples, and dictionaries. We will look at this data type more in detail ahead.

Variables

You have the passengers, but you do not have a mode of commuting; they will have nowhere to go. These passengers would just be folks standing around, waiting for some kind of transportation to pick them up. Similarly, data types cannot function alone. They need to be 'stored' in these vehicles. These special vehicles, or as we programmers refer to as containers, are called 'variables,' and they are the elements that perform the magic for us.

Variables are specialized containers that store a specific value in them and can then be accessed, called, modified, or even removed when the need arises. Every variable that you may create will hold a specific type of data in them. You cannot add more than one type of data within a variable. In other programming languages, you will find that to create a variable, you need to use the keyword 'var' followed by an equals mark '=' and then the value.

In Python, it is a lot easier, as shown below:

```
>   name = "John"

>   age = 33

>   weight = 131.50

>   is_married = True
```

In the above, we have created a variable named 'name' and given it a value of characters. If you recall strings, we have used double quotation marks to let the program know that this is a string. We then created a variable called age. Here, we simply wrote 33, which is an integer as there are no decimal figures following that. You do not need to use quotation marks here at all. Next, we created a variable 'weight' and assigned it a float value. Finally, we created a variable called 'is_married' and assigned it a 'True' bool value. If you were to change the 'T' to 't', the system will not recognize it as a bool and will end up giving an error. Focus on how we used the naming convention for the last variable. We will be ensuring that our variables follow the same naming convention.

You can even create blank variables if you feel like you may need these at a later point in time, or wish to initiate them at no value at the start of the application. For variables with numeric values, you can create a variable with a name of your choosing and assign it a value of zero. Alternatively, you can create an empty string as well by using opening and closing quotation marks only.

```
> empty_variable1 = 0
> empty_variable2 = ""
```

You do not have to necessarily name them like this; you can come up with more meaningful names so that you and any other programmer who may read your code would understand. I have given them these names to ensure anyone can immediately understand their purpose. Now we have learned how to create variables, let's learn how to call them. What's the point of having these variables if we are never going to use them, right? Let's create a new set of variables.
Have a look here:

```
> name = "James"
> age = 43
> height_in_cm = 163
> occupation = "Programmer"
```

I do encourage you to use your own values and play around with variables if you like.
In order for us to call the name variable, we simply need to type the name of the variable.
In order to print that to the console, we will do this:

```
> print(name)
→ James
```

The same goes for age, the height variable, and occupation. But what if we wanted to print them together and not separately? Try running the code below and see what happens:

> print(name age height_in_cm occupation)

→ SyntaxError: invalid syntax

Surprised? Did you end up with this? Here is the reason why that happened. When you were using a single variable, the program knew what variable that was. The minute you added a second, a third, and a fourth variable, it tried to look for something that was written in that manner. Since there wasn't any, it returned with an error that otherwise says: "Umm... Are you sure, Sir? I tried looking everywhere, but I couldn't find this 'name age height_in_cm occupation' element anywhere." All you need to do is add a comma to act as a separator like so:

> print(name, age, height_in_cm, occupation)

→ James 43 163 Programmer

And now, it knew what we were talking about. The system recalled these variables and was successfully able to show us what their values were. But what happens if you try to add two strings together? What if you wish to merge two separate strings and create a third-string as a result?

→ first_name = "John"
→ last_name = "Wick"

To join these two strings into one, we can use the '+' sign. The resulting string will now be called a String Object, and since this is Python we are dealing with, everything within this language is considered as an object, thus the Object-Oriented Programming nature.

> first_name = "John"

> last_name = "Wick"

> first_name + last_name

Here, we did not ask the program to print the two strings. If you wish to print these two instead, simply add the print function and type in the string variables with a + sign in the middle within parentheses. Sounds good, but the result will not be quite what you expect:

```
> first_name = "John"
> last_name = "Wick"
> print(first_name + last_name)
→ JohnWick
```

Hmm. Why do you think that happened? Certainly, we did use a space between the two variables. The problem is that the two strings have combined together, quite literally here, and we did not provide a white space (blank space) after John or before Wick; it will not include that. Even the white space can be a part of a string.

To test it out, add one character of space within the first line of code by tapping on the friendly spacebar after John. Now try running the same command again, and you should see "John Wick" as your result. The process of merging two strings is called concatenation. While you can concatenate as many strings as you like, you cannot concatenate a string and an integer together.

If you really need to do that, you will need to use another technique to convert the integer into a string first and then concatenate the same. To convert an integer, we use the str() function.

```
> text1 = "Zero is equal to"
> text2 = 0
> print(text1 + str(text2))
→ Zero is equal to 0
```

Python reads the codes in a line-by-line method. First, it will read the first line, then the second, then third, and so on. This means we can do a few things beforehand as well, to save some time for ourselves.

> text1 = "Zero is still equal to "

> text2 = str(0)

> print(text1 + text2)

→ Zero is still equal to 0

You may wish to remember this as we will be re-visiting the conversion of values into strings a lot sooner than you might expect. There is one more way through which you can print out both string variables and numeric variables, all at the same time, without the need for '+' signs or conversion. This way is called String Formatting. To create a formatted string, we follow a simple process as shown here:

> print(f"This is where {var 1} will be. Then {var 2}, then {var 3} and so on")

Var 1, 2, and 3 are variables.
You can have as many as you like here. Notice the importance of whitespace. Try not to use the spacebar as much. You might struggle at the start but will eventually get the hang of it. When we start the string, we place the character 'f' to let Python know that this is a formatted string. Here, the curly brackets are performing a part of placeholders. Within these curly brackets, you can recall your variables. One set of curly brackets will be a placeholder for each variable that you would like to call upon. To put this in practical terms, let's look at an example:

> show = "GOT"

> name1 = "Daenerys"

> name2 = "Jon"

> name3 = "Tyrion"

> ```
seasons = 8
```

> ```
print(f'The show called {show} had characters like {name1}, {name2} and {name3} in all {seasons} seasons.')
```

→ The show called GOT had characters like Daenerys, Jon, and Tyrion in all 8 seasons.

If you get an error please make sure you have the latest version of python installed.

While there are other variations to convert integers into strings and concatenate strings together, it is best to learn those which are used throughout the industry as standard. Now, you have seen how to create a variable, recall it, and concatenate the same. Everything sounds perfect, except for one thing; these are predefined values. What if we need input directly from the end-user? How can we possibly know that? Even if we do, where do we store them?

User-Input Values

Suppose we are trying to create an online form. This form will contain simple questions like asking for the user's name, age, city, email address, and so on. There must be some way through which we can allow users to input these values on his/her own and for us to get those back.

We can use the same to print out a message that thanks the users for using the form and that they will be contacted at their email address for further steps. To do that, we will use the input() function. The input function can accept any kind of input. In order to use this function, we will need to provide it with some reference so that the end-user is able to know what he/she is about to fill out. Let us look at a typical example and see how such a form can be created:

> ```
print("Hello and welcome to my interactive tutorial.")
```

> ```
name = input("Your Name: ")
```

> age = int(input("Your age: "))

> city = input("Where do you live? ")

> email = input("Please enter your email address: ")

> print(f"Thank you very much {name}, you will be contacted at {email}.")

→ Hello and welcome to my interactive tutorial.
→ Your Name: Sam
→ Your age: 28

Where do you live? London. Please enter your email address: sam@something.com Thank you very much, Sam, you will be contacted at sam@something.com. In the above, we began by printing a greeting to the user and welcoming them to the tutorial. Next, we created a variable named 'name' and assigned it a value that our user will generously provide us with. In the age, you may have noticed I changed the input to int(), just as we changed integer to string earlier on.

This is because our message within the input parameters is a string value by default, as it is within quotation marks. You will always need to ensure you know what type of value you are after and do the needful, as shown above. Next, we asked for the name of the city and the email address. Now, using a formatted string, we printed out our final message. "Wait! How can we print out something we have yet to receive or know?" I did mention that Python works line by line. The program will start with a greeting, as shown in the output.

Then, it will move to the next line and realize that it must wait for the user to input something and hit enter. This is why the input value has been highlighted by bold and italic fonts here. The program then moves to the next line and waits yet again for the user to put something in and press enter, and this goes on until the final input command is sorted. Now the program has the values stored, it immediately recalls these values and prints them out for the viewer to see in the end. The result was rather pleasing as it gave a personalized message to the user, and we received the information we need.

Everybody walks away, happy! Storing information directly from the user is both essential and, at times necessary. Imagine a game that is based on Python. The game is rather simple, where a ball will jump when you tap the screen. The problem is, your screen isn't responding to the touch at all for some reason. While that happens, the program will either keep the ball running until input is detected or it will just not work at all. We also use input functions to gather information such as login ID and passwords to match with the database, but that is a point that we shall discuss later when we will talk about statements. It is a little more complicated than it sounds at the moment, but once you understand how to use statements; you will be one step closer than ever before to becoming a programmer.

Chapter 5: Numbers in Python

We have briefly looked at numbers in the Data Types chapter. In this chapter, we will go into greater detail on how to use and manipulate numbers in Python. Before we go any further, let's have a brief refresher on the different types of numbers available in Python.

- Integer - These are whole numbers without fractions or decimals.

- Floating Point - These are numbers that have fractional parts expressed with a decimal point.

- Complex - These are complex numbers expressed by using a 'J' or 'j' suffix.

Let's look at a quick example of how these data types can be used.

This program shows how to use number data types in Python.

```
> # This program looks at number data types
> # An int data type
> a=123
> # A float data type
> b=2.23
> # A complex data type
> c=3.14J
> print(a)
> print(b)
> print(c)
```

This program's output will be as follows:

→ 123
→ 2.23
→ 3.14j

There are a variety of functions available in Python to work with numbers. Let's look at a summary of them, after which we will look at each in more detail along with a simple example.

Number Functions

Function	Description
abs()	This returns the absolute value of a number
ceil()	This returns the ceiling value of a number
max()	This returns the largest value in a set of numbers
min()	This returns the smallest value in a set of numbers
pow(x,y)	This returns the power of x to y
sqrt()	This returns the square root of a number
random()	This returns a random value
randrange(start,stop,step)	This returns a random value from a particular range
sin(x)	This returns the sin value of a number
cos(x)	This returns the cosine value of a number
tan(x)	This returns the tangent value of a number

Abs Function

This function is used to return the absolute value of a number. Let's look at an example of this function.

The following program showcases the abs function.

```
>   # This program looks at number functions
>   a=-1.23
```

```
> print(abs(a))
```

This program's output will be as follows:
→ 1.23

Ceil Function

This function is used to return the ceiling value of a number. Let's look at an example of this function. Note that for this program we need to import the 'math' module to use the 'ceil' function.

Example 44: The following program showcases the ceil function.

```
> import math
> # This program looks at number functions
> a=1.23
> print(math.ceil(a))
```

This program's output will be as follows:
2

Max Function

This function is used to return the largest value in a set of numbers. Let's look at an example of this function.

The program below is used to showcase the max function.

```
> # This program looks at number functions
> print(max(3,4,5))
```

This program's output will be as follows:

Min Function

This function is used to return the smallest value in a set of numbers. Let's look at an example of this function.

The following program shows how the min function works.

```
>  # This program looks at number functions
>  print(min(3,4,5))
```

This program's output will be as follows:
3

Pow Function

This function is used to return the value of x to the power of y, where the syntax is 'pow(x,y)'. Let's look at an example of this function.

The following program showcases the pow function.

```
>  # This program looks at number functions
>  print(pow(2,3))
```

This program's output will be as follows:

→ 8

Sqrt Function

This function is used to return the square root of a number. Let's look at an example of this function. Note that for this program we need to import the 'math' module in order to use the 'sqrt' function.

The next program shows how the sqrt function works.

```
>   import math

>   # This program looks at number functions

>   print(math.sqrt(9))
```

This program's output will be as follows:

Random Function

This function is used to simply return a random value. Let's look at an example of this function.

The following program showcases the random function.

```
>   import random

>   # This program looks at number functions

>   print(random.random())
```

The output will differ depending on the random number generated. Also note that for this program we need to use the 'random' Python library.

In command line → pip3 *INSTALL RANDOM*
Then digit

```
>   import random
```

In our case, the program's output is:

→   0.0054600853568235691

Randrange Function

This function is used to return a random value from a particular range. Let's look at an example of this function. Note that we again need to import the 'random' library for this function to work.

This program is used to showcase the random function.

```
> import random
> # This program looks at number functions
> print(random.randrange(1,10,2))
```

The output will differ depending on the random number generated. In our case, the program's output is:

→ 5

Sin Function

This function is used to return the sine value of a number. Let's look at an example of this function.

The following program shows how to use the sin function.

```
> import math
> # This program looks at number functions
> print(math.sin(45))
```

This program's output will be as follows:

→ 0.8509035245341184

Cos Function

This function is used to return the cosine value of a number. Let's look at an example of this function.

This program is used to showcase the cos function.

```
> import math
> # This program looks at number functions
```

```
>   print(math.cos(45))
```

This program's output will be as follows:

→ 0.5253219888177297

Tan Function

This function is used to return the tangent value of a number. Let's look at an example of this function.

The following program shows the use of the tan function.

```
>   import math
>   # This program looks at number functions
>   print(math.tan(45))
```

This program's output will be as follows:
1.6197751905438615

Chapter 6: Operators in Python

1. Arithmetic Operators

These are operators that have the ability to perform mathematical or arithmetic operations that are going to be fundamental or widely used in this programming language, and these operators are in turn subdivided into:

1.1. Sum Operator: its symbol is (+), and its function is to add the values of numerical data. Its syntax is written as follows:

> 6 + 4

→ 10

1.2 Subtract Operator: its symbol is the (-), and its function is to subtract the values of numerical data types. Its syntax can be written like this:

> 4 − 3

→ 1

1.3 Multiplication Operator: Its symbol is (*), and its function are to multiply the values of numerical data types.
Its syntax can be written like this:

> 3 * 2

→ 6

1.4 Division Operator: Its symbol is (/); the result offered by this operator is a real number. Its syntax is written like this:

> 3.5 / 2

→ 1.75

1.5 Module Operator: its symbol is (%); its function is to return the rest of the division between the two operators. In the following example, we have that division 8 is made between 5 that is equal to 1 with 3 of rest, the reason why its module will be 3.
Its syntax is written like this:

> 8 % 5

→ 3

1.6 Exponent Operator: its symbol is (**), and its function is to calculate the exponent between numerical data type values. Its syntax is written like this:

> 3 ** 2

→ 9

1.7 Whole Division Operator: its symbol is (//); in this case, the result it returns is only the whole part.
Its syntax is written like this:

> 3.5 // 2

→ 1.0

However, if integer operators are used, the Python language will determine that it wants the result variable to be an integer as well, this way you would have the following:

> 3 / 2

> 3 // 2

If we want to obtain decimals in this particular case, one option is to make one of our numbers real. For example:

> 3.0 / 2

2. Comparison Operators

The comparison operators are those that will be used to compare values and return; as a result, the True or False response as the case may be, as a result of the condition applied.

2.1 Operator Equal to: its symbol is (= =), its function is to determine if two values are exactly the same.
For example:

> $3 == 3$

→ True

> $5 == 1$

→ False

2.2 Operator Different than: its symbol is (! =); its function is to determine if two values are different and if so, the result will be True.
For example:

> $3 != 4$

→ True

> $3 != 3$

→ False

2.3 Operator Greater than: its symbol is (>); its function is to determine if the value on the left is greater than the value on the right and if so, the result it yields is True. For example:

> $5 > 3$

→ True

> $3 > 8$

→ False

2.4 Operator Less than: its symbol is (<); its function is to determine if the left value is less than the right one, and if so, it gives True result. For example:

> 3 < 5

→ True

> 8 < 3

→ False

2.5 Operator (> =), its function is to determine that the value on the left is greater than the value on the right, if so the result returned is True. For example:

> 8 > = 1

→ True

> 8 > = 8

→ True

> 3 > = 8

→ False

2.6 Operator (< =), its function is to evaluate that the value on its left is less than the one on the right, if so the result returned is True. For example:

> 8 < = 10

→ True

> 8 < = 8

→ True

> 10 < = 8

→ False

3. Logical Operators

Logical operators are the *and, or, not*. Their main function is to check if two or more operators are true or false, and as a result, returns a True or False. It is very common that this type of operator is used in conditionals to return a Boolean by comparing several elements.

Making a parenthesis in operators, we have that the storage of true and false values in Python are of the bool type, and was named thus by the British mathematician George Boole, who created the Boolean algebra. There are only two True and False Boolean values, and it is important to capitalize them because, in lower cases, they are not Boolean but simple phrases.

The semantics or meaning of these operators is similar to their English meaning, for example, if we have the following expression:

X > 0 and x < 8, this will be true if indeed x is greater than zero and less than 8.

In the case of or, we have the following example:

> N=12

> N % 6 = = 0 or n % 8 = = 0

→ True

It will be true if any of the conditions are indeed true, that is, if n is a number divisible by 6 or by 8.

In the case of the logical operator not, what happens is that it denies a Boolean expression, so, if we have, for example:
not (x < y) will be true if x < y is false, that is, if x is greater than y.

Chapter 7: Strings Methods in Python

There will almost certainly be times where you need to manipulate this string or that. Maybe you'll need to get its length, or you'll need to split it or make another string from it. Maybe you'll need to read what character is at x position. Whatever the reason is, the point is that there's a reason.

The reason that we're getting into this is that it opens us up to a broader discussion on the nature of objects that we're going to go more in-depth later on, but in the meantime, we're also going to be covering extremely useful methods that the Python language provides to be used with strings.

Go ahead and create a new file. You can call it whatever you want. My file is going to be named strings.py. Uncreative name, sure, but we're going to be getting creative with strings in this chapter, believe me.

So what is a string, really? Well, we obviously know that a string is a line of text, which goes without saying. But what goes into that?

We've spoken quite a bit in this list about lists. Lists are actually a form of another variable that's largely eschewed in Python programming called an array. An array is a pre-allocated set of data that goes together, in the most basic terms of speaking.

Python comes from and is built upon a language called C. In C, there are actually data types. There are data types in Python, too, but Python saves the user time by setting the data type for the programmer instead of having the programmer declare it.

One of the data types in C was called a char, which was a single character. In terms of computer-speak; there isn't native support for strings. Strings were simply arrays of characters. For example, if one wanted to make a string called "hello", they would have done the following:

→ char hello[6] = { 'h', 'e', 'l', 'l', 'o', '\0' };

Python, in its beautiful habit of maximum abstraction, keeps us from these complexities and lets us just declare:

→ hello = "hello"

The point is that strings, ultimately, are just sets of data. And like any set of data, they can be manipulated. There will be times, too, where we need to manipulate them.

The most simple form of string manipulation is the concept of concatenation. Concatenated strings are strings that are put together to form a new string. Concatenation is super easy - you simply use the + sign to literally add the strings together.

> sentence = "My " + "grandmother " + "baked " + "today."

> print sentence

> # would print "My grandmother baked today."

The first thing to remember when working with string manipulation is that strings, like any set of data, starts counting at 0. So the string "backpack" would count like so:

backpack

01234567

There are a few different things that we can do with this knowledge alone. The first is that we can extract a single letter from it.

Let's say the string "backpack" was stored in a variable called backpack. We could extract the letter "p" from it by typing:

> Letter = backpack[4]

> Print(letter)

→ p

This would extract whatever the character at index 4 was in the string. Here, of course, it's p (start counting from 0, letter b is in position 0).

If we wanted to extract the characters from "b" to "p", we could do the following:

→ substring = backpack[0 : 4]

This would give the variable substring a string equal to the value of backpack's 0 index to 4 index:

backpack
01234567

Substring, thus, would have the value of "backpack". Quite the word. There are a few more things you can do with data sets, and strings specifically, in order to get more specific results.

> backpack[start:4]

would give you all characters from the start to index four, like just before.

> backpack[4:end]

would give you all characters from index 4 to the end.

> backpack[:2]

would give you the first two characters, while backpack[-2:] would give you the last two characters.

> backpack[2:]

would give you everything but the first two characters, while

> backpack[:-2]

would give you everything aside from the last two characters.

However, it goes beyond this simple kind of arithmetic.
String variables also have built-in functions called methods. Most things in Python - or object-oriented languages in general, really - are forms of things called objects. These are essentially variable types that have entire sets of properties associated with them.

Every single string is an instance of the String class, thus making it a string object. The string class contains definitions for methods that every string object can access, as an instance of the String class.

For example, let's create a bit of a heftier string.

> tonguetwister = "Peter Piper picked a peck of pickled peppers"

The string class has a variety of built-in methods you can utilize in order to work with its objects.

Let's take the split method. If you were to type:

> splitList = tonguetwister.split(' ')

It would split the sentence at every space, giving you a list of each word. splitList, thus, would look a bit like this:

['Peter', 'Piper', 'picked', 'a', 'peck', 'of', 'pickled', 'peppers']. Printing splitList[1]

would give you the value 'Piper'.

There's also the count method, which would count the number of a certain character. Typing:

> tonguetwister.lower().count('p')

You would get the number 9.

There's the replace method, which will replace a given string with another. For example, if you typed:

> tonguetwister = tonguetwister.replace("peppers", "potatoes")

tonguetwister would now have the value of "Peter Piper picked a peck of pickled potatoes".

There's the strip, lstrip, and rstrip methods which take either a given character or whitespace off of both sides of the string. This is really useful when you're trying to parse user input. Unstripped user input can lead to unnecessarily large data sets and even buggy code.

The last major one is the join method, which will put a certain character between every character in the string.

> print ("-".join(tonguetwister))

→ "P-e-t-e-r-P-i-p-e-r-p-i-c-k-e-[...]"

There are also various boolean expressions that will return true or false. The starts with(character) and ends with(character) methods are two fantastic examples. If you were to type:

> tonguetwister.startswith("P")

It would ultimately return true. However, if you were to type instead:

> tonguetwister.startswith("H")

It would ultimately return false. These are used for internal evaluation of strings as well as for evaluating user input.

A few other examples are string.isalnum() which will see if all characters in the string are alphanumeric or if there are special characters, string.isalpha() which will see if all characters in the string are alphabetic, string.isdigit() which will check to see if the string is a digit or not, and string.isspace() which will check to see if the string is a space or not.
These are all extremely useful for parsing a given string and making determinations on what to do if the string is or isn't a certain way.

Chapter 8: Program Flow control and If-else, elif Statements in Python

Comparison operators are special operators in Python programming language that evaluate to either True or False state of the condition. Program flow control refers to a way in which a programmer explicitly species the order of execution of program code lines. Normally, flow control involves placing some condition(s) on the program code lines. The most basic form of these conditional statements is the *if statement*. This one is going to provide us with some problems right from the beginning. But knowing a bit about it will help us to get the if else and other control statements to work the way that we want.

To start, the if statement is going to take the input of the user, and compare it to the condition that you set. If the condition is met here, then the code will continue on, usually showing some kind of message that you set up in the code.

However, if the input does not match up with the condition that you set the returned value will be False.

If ... else Flow Control Statements

The "if...else" statement will execute the body of if in the case that the test condition is True. Should the if...else test expression evaluate to false, the body of the else will be executed. Program blocks are denoted <u>by indentation</u>. The if...else provides more maneuverability when placing conditions on the code. The if...else syntax

if test condition:

    Statements

else:

    Statements

A program that checks whether a number is positive or negative
Start IDLE.
Navigate to the File menu and click New Window.
Type the following:

```
number_mine=-56

if(number_mine<0):

        print(number_mine, "The number is negative")

else:

        print(number_mine, "The number is a positive
number")
```

Assignment

Write a Python program that uses if..else statement to perform the following
a.   Given number=9, write a program that tests and displays whether the number is even or odd.
b.   Given marks=76, write a program that tests and displays whether the marks are above pass mark or not bearing in mind that pass mark is 50.
c.   Given number=78, write a program that tests and displays whether the number is even or odd.
d.   Given marks=27, write a program that tests and displays whether the marks are above pass mark or not bearing in mind that pass mark is 50.

Assignment

Write a program that accepts age input from the user, explicitly converts the age into integer data types, then uses if...else flow control to tests whether the person is underage or not, the legal age is 21. Include comments and indentation to improve the readability of the program.

Other follow up work: Write programs in Python using if statement only to perform the following:

1. Given number=7, write a program to test and display only even numbers.

2. Given number1=8, number2=13, write a program to only display if the sum is less than 10.

3. Given count_int=57, write a program that tests if the count is more than 45 and displays, the count is above the recommended number.

4. Given marks=34, write a program that tests if the marks are less than 50 and display the message, the score is below average.

5. Given marks=78, write a program that tests if the marks are more than 50 and display the message, great performance.

6. Given number=88, write a program that tests if the number is an odd number and displays the message, Yes it is an odd number.

7. Given number=24, write a program that tests and displays if the number is even.

8. h. Given number =21, write a program that tests if the number is odd and displays the string, Yes it is an odd number.

Incidental using the If Statement

There are many things that you can do with values and variables, but the ability to compare them is something that will make it much easier for you to try and use Python. It is something that people will be able to do no matter what type of values that they have, and they can make sure that they are doing it in the right way so that their program will appear to be as smooth-running as possible.

To compare your variables is one of the many options that Python offers you, and the best way to do it is through an "if statement."

Now, you can create a new file. This is what you will need to be able to do. Do not forget indentation!

Here is the way that an incidental will look:

```
apples=6

bus = "yellow"

if apples == 0:

    print ("Where are the apples?")

else:

    print ("Did you know that busses are %s?", bus)
```

Run the code through your python program. It will look like this.

→ Did you know that busses are yellow?

The easiest way to understand why the output looked like this is because the apples were not included with the variation. There were not zero apples, and that was something that created a problem with the code. For that reason, it wasn't put in the output because there was no way to do it and no way to make it look again.

To make sure that you are going to be able to use it with a not statement, you can use another if statement in combination with that not.

```
if not apples == 0
```

Now, you can try to run the code again through the program that you created.

Did you know that busses are yellow?

Both of the things that you wrote in the code are included with the statements, and then, you will be able to try different things. If you do not want to write out the not statement, you can simply use the "!"

```
apples=5
if apples!= 0:
....print("How about apples!")
```

When there is an input in your program, such as the number of apples that someone wants or a fact that they have that they can teach you about, the output will look the same. Either they will get a statement about the apples, or they will get a statement about the bus being yellow. If there are no apples that are put into the equation, then you will have the output show up as "Where are the apples"

The conditionals that you use are made up of simple expressions. When you break them down into even smaller pieces, it is easy to understand how they can be used and what you will be able to do with the expressions that you have in the things that you do. It will also give you the chance to be able to show that there is so much more than what you initially had with the variables and values.

if...elif...else Flow Control Statement in Python

Now think of scenarios where we need to evaluate multiple conditions, not just one, not just two but three and more. Think of where you have to choose team member, if not Richard, then Mercy, if not Richard and Mercy then Brian, if not Richard, Mercy, and Brian then Yvonne. Real-life scenarios may involve several choices/conditions that have to be captured when writing a program.

Remember that the elif simply refers to else if and is intended to allow for checking of multiple expressions. The if the block is evaluated first, then elif block(s), before the else block. In this case, the else block is more of a fallback option when all other conditions return false. Important to remember, despite several blocks available in if..elif..else only one block will be executed.. if...elif..else Syntax:

if test expression:

      Body of if

elif test expression:

      Body of elif

else:

      Body of else

Example

Three conditions covered but the only one can execute at a given instance.
Start IDLE.
Navigate to the File menu and click New Window.
Type the following:

```
nNum = 1
if nNum == 0:
        print("Number is zero.")
elif nNum > 0:
        print("Number is a positive.")
else:
        print("Number is a negative.")
```

Nested if Statements in Python

Sometimes it happens that a condition exists but there are more sub-conditions that need to be covered and this leads to a concept known as nesting. The amount of statements to nests is not limited but you should exercise caution as you will realize nesting can lead to user errors when writing code. Nesting can also complicate maintaining of code. The only indentation can help determine the level of nesting.

Example

Start IDLE.
Navigate to the File menu and click New Window.
Type the following:

```
my_charact=str(input("Type a character here either 'a', 'b' or 'c':"))
if (my_charact='a'):
        if(my_charact='a'):
                print("a")
        else if:
                (my_charact='b')
                print("b")
else:
        print("c")
```

Assignment

Write a program that uses the if..else flow control statement to check non-leap year and display either scenario. Include comments and indentation to enhance the readability of the program.

Absolutes

There is a way to create the conditionals so that there is a block of codes that will show you whether or not there is a conditional, and it has something that it can do even if the conditional is not true and cannot be verified with the different things that you do.
That is where the absolute conditionals come into play.
You will need to see whether or not there are different things that you can put in.
Create the variable

apples

Now, you will need to put the input in with the different things that you have created a version of the file that you saved.

> print "What is your age?"

> age = input()

That is the way that you will be able to see how old someone is. But, how exactly does that relate to the number of apples you have?
It doesn't, it just shows you how the variable works so that people can put things in.
You'll create

> apples = input ("What number of apples are there?/n")

That is the easiest part of it and will help you to create the variable that you need for the rest of it.

> if apples == 1:

> print ("I don't know what to do with just one apple!")

You'll get an error though because apples is actually just a string and you need to make it an integer.
Simple:

int(string)

Now it will look like this:

apples = input("What number of apples are there?/n")

apples = int(apples)

if apples == 1:

print ("I don't know what to do with just one apple!")

Put this whole string into your file or change the wording around a bit so that you can figure out what you want to do with it (that is truly great practice for you). When you have put it in, run it through.
The code will work because you created a variable, you added different elements to it, and you allowed for the input of the "apples" in the sequence so that you would be able to show how things worked with it.
This was one of the greatest ways that you could do new things, and it also allowed you the chance to be able to try new things so that you were doing more with them. While you are creating strings of integers, you will need to make sure that you are transforming them into integers instead of simple strings so that you can make sure that they show up and there are no error codes.

Chapter 9: Loops in Python

Before proceeding into loops let's remind once more the *if statement*. The python program will only execute the statements(s) if the test expression is true. The program first evaluates the test expression before executing the statement(s). The program will not execute the statement(s) if the test expression is False. By convention, the body of it is marked by indentation while the first is not indented line signals the end.

for Loop in Python

Indentation is used to separate the body of for loop in Python.

Note: Remember that a simple linear list takes the following syntax:

Variable_name=[values separated by a comma]

Example

Start IDLE.
Navigate to the File menu and click New Window.
Type the following:

numbers=[12,3,18,10,7,2,3,6,1]   #Variable name storing the list

sum=0   #Initialize sum before usage, very important

for i in numbers: #Iterate over the list

    sum=sum+i

    print("The sum is" ,sum)

Assignment

Start IDLE.
Navigate to the File menu and click New Window.
Type the following:
Write a Python program that uses the for loop to sum the following lists.

- → a.   marks=[3, 8,19, 6,18,29,15]
- → b.   ages=[12,17,14,18,11,10,16]
- → c.   mileage=[15,67,89,123,76,83]
- → d.   cups=[7,10,3,5,8,16,13]

range() function in Python

The range function (range()) in Python can help generate numbers. Remember in programming the first item is indexed 0.

Therefore, range(11) will generate numbers from 0 to 10.

Example

Start IDLE.
Navigate to the File menu and click New Window.
Type the following:

```
print(range(7))
```

The output will be 0,1,2,3,4,5,6

Assignment

Without writing and running a Python program what will be the output for:

- • a.   range(16)
- • b.   range(8)
- • c.   range(4)

Assignment

Create a program in Python to iterate through the following list and include the message I listen to (each of the music genre). Use the for loop, len() and range(). Refer to the previous example on syntax.

```
folders=['Rumba', 'House', 'Rock']
```

Using for loop with else

It is possible to include a for loop with else but as an option. The else block will be executed if the items contained in the sequence are exhausted.

Example

Start IDLE.
Navigate to the File menu and click New Window.
Type the following:

```
marks=[12, 15,17]

for i in marks:

        print(i)

else:

        print("No items left")
```

Assignment

Write a Python program that prints all prime numbers between 1 and 50.

While Loop in Python

In Python, the while loop is used to iterate over a block of program code as long as the test condition stays True. The while loop is used in contexts where the user does not know the number of loop cycles that are required; the while loop body is determined through indentation.

Example

Start IDLE.
Navigate to the File menu and click New Window.
Type the following:

```
i=0
while i<6:
        if i==5:
                break
        print(i)
        i=i+1
else:
        print("inside else")
```

Caution: Failing to include the value of the counter will lead to an infinite loop.

Assignment

a.   Write a Python program that utilizes the while flow control statement to display the sum of all odd numbers from 1 to 10.
b.   Write a Python program that employs the while flow control statement to display the sum of all numbers from 11 to 21.
c.   Write a Python program that incorporates a while flow control statement to display the sum of all even numbers from 1 to 10.

Using While loop with else

If the condition is false and no break occurs, a while loop's else part runs.

Example

Start IDLE.
Navigate to the File menu and click New Window.
Type the following:

```
track = 0

while track< 4:

        print("Within the loop")

        track = track + 1

else:

        print("Now within the else segment")
```

Python's break and continue

Let us use a real-life analogy where we have to force a stop on iteration before it evaluates completely. Think of when cracking/breaking passwords using a simple dictionary attack that loops through all possible character combinations, you will want the program immediately it strikes the password searched without having to complete. Again, think of when recovering photos you accidentally deleted using a recovery software, you will want the recovery to stop iterating through files immediately it finds items within the specified range. The break and continue statement in Python works in a similar fashion.

Example

Start IDLE.
Navigate to the File menu and click New Window.
Type the following:

```
for tracker in "bring":
    if tracker == "i":
        break
    print(tracker)
print("The End")
```

Continue statement in Python

When the continue statement is used, the interpreter skips the rest of the code inside a loop for the current iteration only and the loop does not terminate. The loop continues with the next iteration.
The syntax of Python continue
continue

Start IDLE.
Navigate to the File menu and click New Window.
Type the following:

```
for tracker in "bring":
    if tracker == "i":
        continue
    print(tracker)
print("Finished")
```

The output of this program will be:

→ b
→ r
→ n
→ g
→ Finished

Analogy: assume that you are running data recovery software and have specified skip word files (.doc, dox extension). The program will have to continue iterating even after skipping word files.

- ADVANDED PYTHON PROGRAMMING -

by

TechExp Academy

2020 © Copyright

Table of Contents

Chapter 1: Lists in Python

We have learned quite a lot ever since we started with this book. We have gone through operators, we learned about various data types, and we also looked at loops and statements. During all of this, we did mention the word 'list' and represented these with a square bracket instead of curly or round brackets. This chapter will now explore and explain what exactly lists are. By the end of this chapter, we should be familiarized with the core concepts of these and how they are vital to the programming of any kind.

A Look into What Lists Are

Let us go ahead and create an imaginary family that comprises Smith, Mary, their daughter Alicia, and their son Elijah. How would we do that? Begin by creating a variable named family as shown below:

> family = ['Smith', 'Mary', 'Alicia', 'Elijah']

Using the [] brackets, we provided the data to this variable. Now, this specific variable holds more than one name within it. This is where lists come to the rescue. Through listing, we can store as many values within a variable as we like. In this case, we can stick to four only.

If you now use the print command to print 'family', you should see the following:

→ ['Smith', 'Mary', 'Alicia', 'Elijah']

The values or names stored within the brackets are called as items. To call on the item or to check what item is stored on a specific index number, you can use the method we had used earlier in strings.

> print(family[0])

→ Smith

Instead of showing 'S', the complete name was shown. Similarly, if you use the other functions such as the len() function, it would provide you with the length of the list. In this case, it would should you that there are four items in this list. Let us try that out for ourselves.

> print(len(family))

→ 4

You can use the [x:y] where x and y are ranges you can set. This can be helpful if the list you are working on contains hundreds of entries. You can only filter out the ones you would like to view. You can jump straight to the end of the list by using [-1] to see the last entry. The combinations are endless.

Here is a little brain-teaser. Suppose we have numerous numbers in a list, around 100. They are not listed chronologically and we do not have time to scroll through each one of them. We need to find out which of these numbers is the highest. How can we do that?

This is where lists, loops, and if statements come together. How? Let us look into it right away:

```
numbers = [312, 1434, 68764, 4627, 84, 470, 9047, 98463,
389, 2]

high = numbers[0]

for number in numbers:

        if number > high:

                high = number

        print(f"The highest number is {high}")
```

→ The highest number is 98463

Time to put our thinking cap on and see what just happened.

We started out by providing some random numbers. One of these was surely the highest. We created a variable and assigned it the first item of the list of numbers as its value. We do not know whether this item holds the highest value.

Moving ahead, we initiated a 'for' loop where we created a loop variable called number. This number would iterate each value from numbers. We used an 'if' statement to tell the interpreter that if the loop variable 'number' is greater than our current set highest number 'high', it should immediately replace that with the value it holds. Once the program was run, Python sees we assigned 'high' a value of the first item which is 312. Once the loop and if statement begins, Python analyzes if the first item is greater than the value of the variable 'high'. Surely, 312 is not greater than 312 itself. The loop does not alter the value and ends. Now, the 'for' loop restarts, this time with the second item value. Now, the value has changed. This time around, when the 'if' statement is executed, Python sees that our variable has a lower value than the one it is currently working on. 312 is far less than in 1434. Therefore, it executes the code within the statement and replaces the value of our variable to the newly found higher value. This process will continue until all values are cross-checked and finally the largest value is maintained. Then, only the largest value will be printed for us.

2-D Lists

In Python, we have another kind of list that is called the two-dimensional list. If you are someone who is willing to master data sciences or machine learning, you will need to use these quite a lot. The 2-D list is quite a powerful tool. Generally, when it comes to maths, we have what is called matrixes. These are arrays of numbers formed in a rectangular form within large brackets.

Unlike your regular lists, these contain rows and columns of values and data as shown here:

matrix = [

    [19, 11, 91],

    [41, 25, 54],

    [86, 28, 21]]

In an easier way, imagine this as a list that contains a number of lists inside. As illustrated above, each row is now acting as a separate list. Had this been a regular list, we could have printed a value using the index number. How do you suppose we can have the console print out the value of our first item within the first list?

```
> print(matrix[0][0])
```

Using the above, you can now command interpreter to only print out the first value stored within the first list. The first zero within the first [] tells the interpreter the number of list to access. Following that is the second bracket set which further directs the search to the index number of the item. In this case, we were aiming to print out 19 and thus, 19 will be our result.

Take a moment and try to print out 25, 21, and 86 separately. If you were able to do this, good job.

You can change the values of the items within the list. If you know the location of the said item, you can use the name of the variable followed by the [x][y] position of the item. Assign a new number by using the single equal to mark and the value you wish for it to have.

The 2-D lists are normally used for slightly advanced programming where you need to juggle quite a lot of values and data types. However, it is best to keep these in mind as you never know when you may actually need to use them.

List Methods

Somewhere in the start, we learned about something called methods. To start off, let us go back to the PyCharm and create our own list of random numbers. Let's use the following number sequence:

```
> numbers = [11, 22, 33, 44, 55, 66, 77]
```

We are not going to print this out to our console. Instead, we would like to see what possible methods are available for us to use. In the next line, type the name of the variable followed by the dot operator "." to access the methods.

Let us type the append method.

```
> numbers.append(10)
```

The append method allows us to add an entry or a value to the list under the selected variable. Go ahead and add any number of your choice. Done? Now try and print the variable named 'numbers' and see what happens.

You should be able to see a number added at the end of the list. Good, but what if you don't wish to add a number at the end? What if you want it to be somewhere close to the start?

To do that, we need a method called insert.

numbers.insert()

In order for us to execute this properly, we will need to first provide this method with the index position where we wish for the new number to be added. If you wish to add it to the start, use zero, or if you wish to add it to any other index, use that number. Follow this number by a comma and the number itself.

Now, if you print the numbers variable, you should be able to see the new number added exactly where you wanted.

> numbers.insert(2, 20)

> print(numbers)

→ [11, 22, 20, 33, 44, 55, 66, 77]

Similarly, you can use a method called remove to delete any number you wish to be removed from the list. When using the remove method, do note that it will only remove the number where it first occurred. It will not remove the same number which might have repeated later on within the same list as shown here:

> numbers = [11, 22, 33, 44, 55, 66, 77, 37, 77]

> numbers.remove(77)

> print(numbers)

→ [11, 22, 33, 44, 55, 66, 37, 77]

For any given reason, if you decide you no longer require the list content, you can use the clear command. This command does not require you to pass any object within the parentheses.

> numbers.clear()

Using another method, you can check on the index number of a specific value's first occurrence.

→ numbers.index(44)

If you run the above, you will get '3' as a result. Why? The index position of three contains the number 44 in the list we used earlier. If you put in a value that is not within the defined list values, you will end up with an error as shown here:

> print(numbers.index(120))

→ ValueError: 120 is not in list

There is another useful method that helps you quite a lot when you are dealing with a bunch of numbers of other data types. In case you are not too sure and you wish to find out whether a specific number exists within a list, you can use the 'in' operator as shown:

> numbers = [11, 22, 33, 44, 55, 66, 77, 37, 77]

> print(43 in numbers)

What do you think the result will be? An error? You might be wrong. This is where the result will show 'False'. This is a boolean value and is indicating that the number we wanted to search for does not exist in our list. If the number did exist, the return boolean value would have been 'True'.

Let us assume that we have a large number of items in the list and we wish to find out just how many times a specific number is being used or repeated within the said list. There is a way we can command Python to do it for us. This is where you will use the 'count' method.

In our own list above, we have two occurrences where the number 77 is used. Let us see how we can use this method to find out both instances.

```
>   print(numbers.count(77))
```

The result will now state '2' as our result. Go ahead and add random numbers to the list with a few repeating ones. Use the count method to find out the number of occurrences and see how the command works for you. The more you practice, the more you will remember.

Now we have seen how to locate, change, add, clear, and count the items in the list. What if we wish to sort the entire list in ascending or descending order? Can we do that?

With the help of the sort method, you can actually have that carried out. The sort() method by default will only sort the data into ascending order. If you try and access the method within a print command, the console will show 'none' as your return. To have this done correctly, always use the sort method before or after the print command. To reverse the order, use the reverse() method. This method, just like the sort() method, does not require you to pass any object within the brackets.

Tuples

In Python, we use lists to store various values and these values can be accessed, changed, modified, or removed at will, whenever we like. That certainly might not be the best thing to know if you intend to use data that is essential in nature. To overcome that, there is a kind of list that will store the data for you but after that, no additional modification will be carried out, even if you try and do it accidentally or intentionally. These are called tuples.

Tuples are a form of list which are very important to know when it comes to Python. Unlike the square bracket representation for lists, these are represented by parentheses ().

```
>   numbers = (19, 21, 28, 10, 11)
```

Tuples are known as immutable items. This is because of the fact that you cannot mutate or modify them. Let us deliberately try and modify the value to see what happens.

As soon as you type in the dot operator to access append, remove, and other similar methods, you should see this instead:

```
numbers = (19, 21, 28, 10, 11)
numbers.
        count(self, x)                              tuple
        index(self, x, start, end)                  tuple
        __add__(self, x)                            tuple
        __annotations__                             object
        __class__                                   object
        __contains__(self, x)                       tuple
        __delattr__(self, name)                     object
        __dict__                                    object
        __dir__(self)                               object
        __eq__(self, o)                             object
        __format__(self, format_spec)               object
```

Tuples simply do not have these options anymore. That is because you are trying to modify a value that is secure and locked by Python. You can try another method to see if you can forcefully change the value by doing this:

> numbers = (19, 21, 28, 10, 11)

> numbers[0] = 10

> print(numbers)

> numbers[0] = 10

→ TypeError: 'tuple' object does not support item assignment

See how the error came up? The program cannot carry out this change of value, nor can it append the value in any way.

While most of the time you will be working with lists, tuples come in handy to ensure you store values that you know you don't wish to change accidentally in the future. Think of a shape that you wish to create and maintain throughout the game or website as uniform. You can always call on the values of a tuple and use the values when and where needed.

The only way these values might be changed is if you purposely or unintentionally overwrite them. For example, you had written the values of a tuple within the code and move on hundreds of lines ahead. At this point, you might have forgotten about the earlier values or the fact that you wrote these values previously. You start writing new values by using exactly the same name and start storing new values within them. This is where Python will allow you to overwrite the previously stored values without providing you any errors when you run the program.

The reason that happens is because Python understands that you may wish for a value to change later on and then stay the same for a while until you need to change them yet again. When you execute the program, the initially stored values will continue to remain in use right up until the point where you want them to be changed. In order to do that, you can simply do the following:

```
> numbers = (1, 2, 3, 4, 5)

> print(numbers)

→ (1, 2, 3, 4, 5)

> numbers = (6, 7, 8, 9, 10)

> print(numbers)

→ (6, 7, 8, 9, 10)
```

The number values have changed without the program screaming back at us with an error. As long as you know and you do this change on purpose, there is absolutely nothing to worry about. However, should you start typing the same tuple and are about to rewrite it, you will be notified by PyCharm about the existence of the same tuple stored before. Can you guess how? Go ahead and try writing the above example in PyCharm and see how you are notified.

PyCharm will highlight the name of the tuple for you, and that is an indication that you have already used the same number before. If this was the first occurrence, PyCharm will not highlight the name or the values for you at all.

Unpacking

Since we just discussed tuples, it is essential to know about a feature that has further simplified the use of tuples for us. Unpacking is of great help and is quite useful, too. Suppose you have a few values stored in a tuple and you wish to assign each one of them to another variable individually. There are two ways you can do that. Let us look at the first way of doing so and then we will look at the use of unpacking for comparison.

First method:

```
>  ages = (25, 30, 35, 40)
>  Drake = ages[0]
>  Emma = ages[1]
>  Sully = ages[2]
```

If you print these values now, you will see the ages accordingly. This means that the values stored within these individual variables were successfully taken from the tuple as we wanted. However, this was a little longer. What if we can do all of that in just one line?

Second method:

```
>  ages = (25, 30, 35, 40)
>  Drake, Emma, Sully, Sam = ages
```

Now, this looks much more interesting. Instead of using a number of lines, we got the same job done within the same line. Each individual variable still received the same age as the first method and each can be called upon to do exactly the same thing. This is how unpacking can work miracles for us. It saves you time and effort and allows us to maintain a clean, clear, and readable code for reference.

With that said, it is now time for us to be introduced to one of the most important elements within Python that is used both by beginners and experts almost every single time.

Python Dictionaries

Dictionaris are a data structure that enables to memorize the data in key-value couple.

AlumniAge = {'Andrea': 23, 'John': 28}

There are two keys- 'Andrea' and 'John'. The value associated with key 'Andrea' is '23' which is his age. The value associated with key John is '28' which is his age.

These are the simple things you should always remember while using a Python dictionary.

Key in the dictionary are unique. Keys are immutable. You can always eliminate and add a new key. But, you can not update the key.

There are times you will come across certain information that is unique and holds a key value. Let us assume that you have to design software that can store information about customers or clients. This information may include and is not limited to names, numbers, emails, physical addresses, and so on. This is where dictionaries will come into play.

If you had thought that a dictionary in Python would be like your everyday dictionary for languages we speak, you might not have been completely wrong here. There is a similarity that we can see in these dictionaries. Every single entry that is made is unique. If an entry tries to replicate itself or if you try to store the same value again, you will be presented with an error.

So how exactly do we use dictionaries? For that, let us switch back to our new best friend, the PyCharm, and start typing a little.

Come up with an imaginary person's name, email address, age, and phone number. Don't start assigning these yet, as we would like to use the dictionary here to do the same. Ready? Okay, let us begin.

```
→  user_one = {    #Dictionaries are represented by {}
→      'name': 'Sam',
→      'age': 40,
→      'phone': 123456789,
→      'married': False
→  }
```

We have entered some information about a virtual character named Sam. You can use the print command and run the dictionary named 'user_one' and the system will print out these values for you.

For dictionaries, we use the colon: sign between values. The object name is placed in a string followed by the colon sign. After that, we use either a string, a number (integer or float), or a boolean value. You can use these to assign every object with its unique key pair. In case you are confused, the key pair is just another way of saying the value that is assigned to the object. For example, the key pair for 'name' is 'Sam'.

Now, let us try and see what happens if we add another 'married' value. As soon as you are done typing, the system will highlight it straight away. Note that you can still type in the new value and the system will continue to function. However, the value it will use will be the latest value it can find.

This means that if you initially set the value for married to False and later change it to True, it will only display True.

user_one = {   #Dictionaries are represented by {}
```
        'name': 'Sam',
        'age': 40,
        'phone': 123456789,
        'married': False,
        'married': True}
```

print(user_one['married'])

```
→  True
```

When it comes to calling values from the dictionary, we use the name of the string instead of the index number. If you try and run the index number zero, you will be presented with a 'KeyError: 0' in the traceback. Can you guess why that happens?

Dictionaries store values that are unique. If you use a number or a name that does not exist within the defined dictionary, you will always end up with an error. You will need to know the exact name or value of the information you are trying to access.

Similarly, if you try to access 'Phone' instead of 'phone', you will get the same error as Python is case-sensitive and will not identify the former as an existing value.

Dictionaries can be updated easily should the situation call for it. Let us assume that we got the wrong phone number for our client stored in 'user_one', we can simply use the following procedure to update the entry right away:

```
> user_one['phone'] = 345678910
> print(user_one['phone'])
```

You should now be able to see the new number we have stored. There's one little thing you may have noticed right about now when you did this. See the crazy wiggly lines which have appeared? These are here to suggest you rewrite the value instead of updating it separately to keep the code clean. PyCharm will continue to do this every now and then where it feels like you are causing the code to grow complicated. There is no reason for you to panic if you see these lines. However, if the lines are red in color, something is surely wrong and you may need to check on that.

Similarly, if you wish to add new key information to your dictionary, you can do so easily using almost the same process as shown here:

```
> user_one['profession'] = 'programmer'
```

It is that easy! Try and print out the information now and you should be able to see this along with all previous entries available to you.

Lastly, you can use a method called 'get' to stop the program from coming back with an error in case you or your program user enters a wrong or a missing value when calling upon a dictionary. You can also assign it a default value like a symbol to notify yourself or the user that this value does not exist or is not identifiable by the program itself. Here is a little example where the user has tried to find out information about 'kids'. We have provided it with a default value of 'invalid':

```
> print(user_one.get('kids', 'invalid'))
```

If you run this through, you will be presented with a result that shows an object named 'invalid'. We will make use of this feature in a more meaningful way in our test.

- clear(): Remove all items from the dictionary.
- copy(): Return a shallow copy of the dictionary.
- fromkeys(seq[, v]): Return a new dictionary with keys from seq and value equal to v (defaults to None).
- get(key[,d]): Return the value of key. If the key does not exist, return d (defaults to None).
- items(): Return a new view of the dictionary's items (key, value).
- keys(): Return a new view of the dictionary's keys.
- pop(key[,d]): Remove the item with a key and return its value or return 'd' if the key is not found. If d is not provided and the key is not found, raises KeyError.
- popitem(): Remove and return an arbitrary item (key, value). Raises KeyError if the dictionary is empty.
- setdefault(key[,d]): If the key is in the dictionary, return its value. If not, insert key with a value of d and return d (defaults to None).
- update([other]): Update the dictionary with the key/value pairs from other, overwriting existing keys.

values(): Return a new view of the dictionary's value

There are also some embedded functions that can be used with a dictionary.
- all(): Return True if all keys of the dictionary are true (or if the dictionary is empty).
- any(): Return True if any key of the dictionary is true. If the dictionary is empty, return False.
- len(): Return the length (the number of items) in the dictionary.
- cmp(): Compares items of two dictionaries.
- sorted(): Return a new sorted list of keys in the dictionary.

Chapter 2: Deep Dive on Python Tuples

Tuple in Python

A tuple is like a list, but we cannot change elements in a tuple. This is one of the main differences that we will see with the tuple versus the list when we work in the Python language, and other coding languages as well.

For example, when we work with a list, we have different items that are available in the code, and we are able to change them, even when we execute the code. This is a benefit of the list; if you want to be able to change up some of the items and not have them set there for a long time, then you would want to work with the list.

On the other hand, the tuple is going to look similar in syntax to what we see with a list. But the elements are not changeable at all. If you want to work with an option that will keep the elements the same all of the time, once you are done putting them in place, then the tuple is going to be the best option for you.

Example:
Start IDLE.
Navigate to the File menu and click New Window.
Type the following:

```
>  tuple_mine = (21, 12, 31)
>  print(tuple_mine)
>  tuple_mine = (31, "Green", 4.7)
>  print(tuple_mine)
```

Accessing Python Tuple Elements

```
>  tuple_mine=['t','r','o','g','r','a','m']
>  print(tuple_mine[1])#output:'r'
>  print(tuple_mine[3])#output:'g'
```

Negative Indexing

Just like lists, tuples can also be indexed negatively.
Like lists, -1 refers to the last element on the list and -2 refer to the second last element.
Example:

Start IDLE.
Navigate to the File menu and click New Window.
Type the following:

```
> tuple_mine=['t','r','o','g','r','a','m']
> print(tuple_mine [-2]) #the output will be 'a'
```

Slicing

The slicing operator, the full colon is used to access a range of items in a tuple.
Example:

Start IDLE.
Navigate to the File menu and click New Window.
Type the following:

```
> tuple_mine=['t','r','o','g','r','a','m']
> print(tuple_mine [2:5]) #Output: 'o','g','r','a'
> print(tuple_mine[:-4]) #'g','r','a','m'
```

Important
Tuple elements are immutable meaning they cannot be changed. However, we can combine elements in a tuple using +(concatenation operator). We can also repeat elements in a tuple using the * operator, just like lists.
Example:
Start IDLE.
Navigate to the File menu and click New Window.
Type the following:

> print((7, 45, 13) + (17, 25, 76))

> print(("Several",) * 4)

Since we cannot change elements in a tuple, we cannot delete the elements too. However, removing the full tuple can be attained using the keyword del.

Example:
Start IDLE.
Navigate to the File menu and click New Window.
Type the following:

```
→  t_mine=['t','k','q','v','y','c','d']
→  del t_mine
```

Inbuilt Python Functions with Tuple

Method	Description	Method	Description
enumerate()	Return an enumerate object. It contains the index and value of all the items of a tuple as pairs.	tuple()	Convert an iterable to a tuple.
sorted()	Take elements in the tuple and return a new sorted list(does not sort the tuple itself)	max()	Return the largest element in the tuple.
all()	Return True if all elements of the tuple are true(or if the tuple is empty).	sum()	Return the sum of all elements in the tuple.
len()	Return the length(the number of items) in the tu-	min()	Return the smallest item in the tuple.

	ple.		
		any()	Return True if any element of the tuple is true. If the tuple is empty, return False.

Escape Sequences in Python

The escape sequences enable us to format our output to enhance clarity to the human user. A program will still run successfully without using escape sequences, but the output will be highly confusing to the human user. Writing and displaying output in expected output is part of good programming practices. The following are commonly used escape sequences.

Method	Description	Method	Description
\n	ASCII Linefeed	\b	ASCII Backspace
\"	Double quote	\\	Backslash
\f	ASCII Formfeed	\a	ASCII Bell
\newline	Backslash and newline ignored	\'	Single quote
\r	ASCII Carriage Return	\t	ASCII Horizontal Tab
\v	ASCII Vertical Tab	\ooo	Character with octal value ooo
\xHH	The character with hexadecimal value HH		

Examples:
Start your IDLE.
Navigate to the File menu and click New Window.
Type the following:

> → print("D:\\Lessons\\Programming")
> → print("Prints\n in two lines")

Integers, floating-point, and complex numbers are supported in Python. There are integers, floating and complex classes that help convert different number data types. The presence or absence of a decimal point separates integers and floating points. For instance, 4 is an integer while 4.0 is a floating-point number. Programmers often need to convert decimal numbers into octal, hexadecimal, and binary forms. We can represent binary, hexadecimal, and octal systems in Python by simply placing a prefix to the particular number. Sometimes referred to as coercion, type conversion allows us to change one type of number into another.

Inbuilt functions such as int() allow us to convert data types directly. The same functions can be used to convert from strings. We create a list in Python by placing items called elements inside square brackets separated by commas. In programming and Python specifically, the first time is always indexed zero. For a list of five items, we will access them from index0 to index4. Failure to access the items in a list in this manner will create an index error.

Chapter 3: Sets in Python

Sets

A Python set is an unordered collection containing unique items. While sets are mutable, meaning that we can add items to them and remove items, each item must itself be immutable. One important thing about sets is that you cannot have any duplicate items in them. We usually use sets to do mathematical operations, like union, intersection, complement, and difference.

Unlike a sequence type, a set type doesn't provide any slicing or indexing operations. The values do not have any keys associated with them as the dictionaries do. Python contains two types of sets – mutable set and immutable frozenset. Each type of set is created the same way with a set of curly braces containing values each separated by a comma. An empty set using the curly braces cannot be created through as a={} will create a dictionary! Empty sets are created by using a=set() or a=frozenset().

Below are the methods and the operations for sets:

len(s)
will return how many elements are in (s)

s.copy()
will return a shallow copy of (s)
s.difference(t)
will return a set containing the items in (s) but not (t)

s.intersection(t)
will return a set containing the items in both (s) and (t)

s.isdisjoint(t)
will return True if there are no common items between (s) and (t)

s.issubset(t)
will return True if the contents of (s) are in (t)
s.issuperset(t)
will return True if the contents of (t) are in (s)

s.symmetric_difference

will return a set of the items from (s) or (t) but not those in both

s.union(t)
will return a set containing the items in (s) or (t)

The parameter of (t) may be any object in Python that will support iteration and any method that is available to the objects set and frozenset. Be aware, if you need to use an operator version of these methods, you must set the arguments – the methods themselves will accept any type that is iterable. Below are the operator versions of the methods:

Operator Methods:

> s.difference(t)

> s – t – t2 - ....

> s.issubset(t)

> s<=t

> s<(t(s!=t)

> s.issuperset(t)

> s>=t

> s>t(s!=t)

> s.symmetric_difference(t)

> s^t

> s.union(t)

> s | t1 | t2 | ....

For the mutable set objects, there are some extra methods:
s.add(item)

---

will add the item to (s) – if the item is already there, nothing will happen

s.clear()
all items are removed from (s)

s.difference_update(t)
will remove items in (s) that are also in (t)

s.discard(item)
will remove the specified item from (s)

s.intersection_update(t)
will remove items from (s) that do not appear in the intersection of (s) and (t)

s.pop()
will return and remove a given arbitrary item from (s)

s.remove(item)
will remove the specified item from (s)

s.symmetric_difference_update(t)

will remove items from (S) that cannot be found in the symmetric difference of (s) and (t)

s.update(t)
will add the items from iterable (t) to (s)

The next example will show you a couple of set operations and the results:

```
>  s1={'ab', 3, 4, (5, 6)}

>  s2={'ab', 7, (7, 6)}

>  print(s1-s2) # identical to s1.difference (s2)
```

Running this code will return:

{(5, 6), 3, 4}

```
> print(s1.intersection(s2))
```

Running this code will return:
→ {'ab'}

```
> print(s1.union(s2))
```

Running this code will return:

→ { 'ab', 3, 4, 7, (5, 6), (7, 6)}

The set object is not bothered if its members are of different types; they just have to all be immutable. If you were to try using mutable objects, like a list or a dictionary, in your set, an unhashable type error will be thrown. All hashable types have got a hash value that remains the same through the entire lifecycle of the instance and all of the built-in immutable types are of a hashable type. The built-in mutable types are not of a hashable type so cannot be included in sets as elements or as dictionary keys.

Note, when the s1 and the s2 union is printed, there is just one 'ab' value. Remember, sets do not include any duplicates. As well as the built-in methods, there are several other operations that can be performed on a set. For example, if you wanted to test a set for membership, you would do the following:

→ print('ab' in s1)

Running this code will return:
True

→ print('ab' not in s1)
Running this code will return:

False

Looping through the elements of a set would look like this:

```
>   for element in s1:
        print(element)
```

```
→   (5, 6)
→   ab
→   3
→   4
```

Immutable Sets

As mentioned earlier, Python has an immutable set type known as frozenset. It works much the same as set does but with one difference – you cannot use any operation or method that can change values, such as the clear() or add() methods. Immutability can be quite useful in a number of ways. For example, a normal set is mutable, which means it is not hashable and that, in turn, means that it cannot be a member of another set. On the other hand, frozenset can be a member of another set because it is immutable. Plus, this immutability of frozenset means that it can be used as a dictionary key, like this:

```
→   fs1 = frozenset(s1)
→   fs2 = frozenset(s2)
→   print({fs1: 'fs1', fs2: 'fs2'})
```

Running this code will return:

```
→   { frozenset({(5, 6), 3, 4, 'ab'}): 'fs1', frozenset({(7, 6),
    'ab', 7}); 'fs2'}
```

Modules for Algorithms and Data Structures

As well as these built-in types, Python has a few modules that can be used for extending those built-in functions and types. Much of the time, the modules may be more efficient and may offer advantages in programming terms that let us make our code simpler.

Up to now, we have examined the built-in types for dictionaries, lists, sets, and strings. These are often termed as ADTs or abstract data types and may be considered to be mathematical specifications for the operations that may be performed on the data. Their behavior is what defines them, not their implementation. Besides the ADTs we already looked at, there are a few Python libraries that include extensions for the built-in datatypes and we'll be looking at these next.

Collections

The module called collections contains alternatives for these data types, high-performance specialized alternatives, in addition to a utility function for the creation of named tuples. Below are the datatypes, the operations, and what they do:

namedtuple()

will create a tuple subclass with the given fields

deque

provides lists that have fast appends and that pop at either end

ChainMap

a class much like a dictionary that creates a single view showing multiple mappings

Counter

a dictionary subclass used for counting the hashable objects

OrderedDict

a dictionary subclass that will remember the order of entries

defaultdict

a dictionary subclass that will call a function that can supply the missing values

- UserDict
- UserList
- UserString

These datatypes are used as wrappers for the base class that underlies each one. We don't tend to use these very much now but subclass the base classes instead. They can, however, be used for accessing the underlying objects as attributes.

Deques

A deque (pronounced deck) is a double-ended queue. These objects are similar to lists with support for appends that are thread-safe and memory-efficient. A deque is mutable and has support for some of the list operations, like indexing. You can assign a deque by index, but you cannot slice one directly. For example, if you tried dq[1:2], you would get a Type Error; there is a way to do it though and we'll look at that later.

The biggest advantage of using a deque and not a list is that it is much faster to insert an item at the start of a deque than it is at the start of a list. That said, it isn't always faster to insert items at the end of a deque. All deques are thread-safe and we can serialize them using a module called pickle.

Perhaps one of the most useful ways to use a deque is for the population and consummation of items. Population and consummation of deque items happen sequentially from each end.

```
> from collections import deque

> dq = deque('abc') #the deque is created

> dq.append('d') #value 'd' is added on the right

> dq.appendleft('z') # value 'z' is added on the left

> dq.extend('efg') # multiple items are added on the right

> dq.extendleft('yxw') # multiple items are added on the left

> print(dq)
```

Running this code will return:
→ deque(['w', 'x', 'y', 'z', 'a', 'b', 'c', 'd', 'e', 'f', 'g'])

To consume items in our deque, we can use the pop() and popleft() methods:

```
> print(dq.pop()) #item on the right is returned and removed
```

Running this code will return:
→ 'g'

> print(dq.popleft()) # item on the left is returned and removed

Running this code will return:
→ 'w'

→ print(dq)
Running this code will return:
→ deque(['x', 'y', 'z', 'a', 'b', 'c', 'd', 'e', 'f'])

We also have a method called rotate(n) which moves and rotates any item of n steps right for positive n integer values and left for negative n integer values, with positive integers used for the argument:

> dq.rotate(2) #all the items are moved 2 steps to the right

> print(dq)

Running this code will return:
→ deque(['e', 'f', 'x', 'y', 'z', 'a', 'b', 'c', 'd'])

> dq.rotate(-2) # all the items are moved 2 steps to the left

> print(dq)

Running this code will return:
→ deque(['x', 'y', 'z', 'a', 'b', 'c', 'd', 'e', 'f'])

We can also use these pop() and rotate() methods to delete elements too. The way to return a slice as a list is this:

```
> print(dq)
```

Running this code will return:

```
> deque(['x', 'y', 'z', 'a', 'b', 'c', 'd', 'e', 'f'])
```

```
> import itertools
> print(list(itertools.islice(dq, 3, 9)))
```

Running this code will return:
→ ['a', 'b', 'c', 'd', 'e', 'f']

The itertools.islice() method is the same as a slice on a list, but instead of the argument being a list, it is an iterable and selected values are returned as a list, using start and stop indices.

One of the more useful deque features is that a maxlen parameter is supported as optional and used for restricting the deque size. This is ideal for circular buffers, a type of data structure that is fixed in size and is connected end-to-end. These tend to be used when data streams are to be buffered. Here is an example:

```
→ dq2=deque([],maxlen=3)
  for i in range(6):
      dq2.append(i)
      print(dq2)
```

We have populated from the right and consumed from the left. As soon as the buffer has been filled, the older values are consumed, and new values come in from the right.

Chapter 4: Functions in Python

Creating and calling a function is easy. The primary purpose of a function is to allow you to organize, simplify, and modularize your code. Whenever you have a set of code that you will need to execute in sequence from time to time, defining a function for that set of code will save you time and space in your program. Instead of repeatedly typing code or even copy pasting, you simply define a function.

We began with almost no prior knowledge about Python except for a clue that it was some kind of programming language that is in great demand these days. Now, look at you; creating simple programs, executing codes, and fixing small-scale problems on your own. Not bad at all! However, learning always comes to a point where things can get rather trickier.

In quite a similar fashion, Functions are docile looking things; you call them when you need to get something done. But did you know that these functions have so much going on at the back? Imagine every function as a mini-program. It is also written by programmers like us to carry out specific things without having us to write lines and lines of codes. You only do it once, save it as a function, and then just call the function where it is applicable or needed.

The time has come for us to dive into a complex world of functions where we don't just learn how to use them effectively, but we also look into what goes on behind these functions, and how we can come up with our very own personalized function. This will be slightly challenging, but I promise, there are more references that you will enjoy to keep the momentum going.

How to define and call function?

To start, we need to take a look at how we are able to define our own functions in this language. The function in Python is going to be defined when we use the statement of "def" and then follow it with a function name and some parentheses in place as well. This lets the compiler know that you are defining a function, and which function you would like to define at this time as well. There are going to be a few rules in place when it comes to defining one of these functions though, and it is important to do these in the proper manner to ensure your code acts in the way that you would like. Some of the Python rules that we need to follow for defining these functions will include:

1. Any of the arguments or input parameters that you would like to use have to be placed within the parentheses so that the compiler knows what is going on.

2. The function first statement is something that can be an optional statement something like a documentation string that goes with your function if needed.

3. The code that is found within all of the functions that we are working with needs to start out with a colon, and then we need to indent it as well.

4. The statement return that we get, or the expression, will need to exit a function at this time. We can then have the option of passing back a value to the caller. A return statement that doesn't have an argument with it is going to give us the same return as none.

Before we get too familiar with some of the work that can be done with these Python functions, we need to take some time to understand the rules of indentation when we are declaring these functions in Python. The same kinds of rules are going to be applicable to some of the other elements of Python as well, such as declaring conditions, variables, and loops, so learning how this work can be important here.

You will find that Python is going to follow a particular style when it comes to indentation to help define the code because the functions in this language are not going to have any explicit begin or end like the curly braces in order languages to help indicate the start and the stop for that function. This is why we are going to rely on the indentation instead. When we work with the proper kind of indentation here, we are able to really see some good results and ensure that the compiler is going to know when the function is being used.

Understanding Functions Better

Functions are like containers that store lines and lines of codes within themselves, just like a variable that contains one specific value. There are two types of functions we get to deal with within Python. The first ones are built-in or predefined, the others are custommade or user-created functions.

Either way, each function has a specific task that it can carry out. The code that is written before creating any function is what gives that function identity and a task. Now, the function knows what it needs to do whenever it is called in.

When we began our journey, we wrote "I made it!" on the console as our first program? We used our first function there as well: the print() function. Functions are generally identified by parentheses that follow the name of the function. Within these parentheses, we pass arguments called parameters. Some functions accept a certain kind of parenthesis while others accept different ones.

Let us look a little deeper and see how functions greatly help us reduce our work and better organize our codes. Imagine, we have a program that runs during live streaming of an event. The purpose of the program is to provide our users with a customized greeting. Imagine just how many times you would need to write the same code again and again if there were quite a few users who decide to join your stream. With functions, you can cut down on your own work easily.

In order for us to create a function, we first need to 'define' the same. That is where a keyword called 'def' comes along. When you start typing 'def' Python immediately knows you are about to define a function. You will see the color of the three letters change to orange (if using PyCharm as your IDE). That is another sign of confirmation that Python knows what you are about to do.

→ def say_hi():

Here, say_hi is the name I have decided to go with, you can choose any that you prefer. Remember, keep your name descriptive so that it is understandable and easy to read for anyone. After you have named your function, follow it up with parentheses. Lastly, add the friendly old colon to let Python know we are about to add a block of code. Press enter to start a new indented line.

Now, we shall print out two statements for every user who will join the stream.

print("Hello there!")

print('Welcome to My Live Stream!')

After this, give two lines of space to take away those wiggly lines that appear the minute you start typing something else. Now, to have this printed out easily, just call the function by typing its name and run the program. In our case, it would be:

```
>   say_hi()
```

→ Hello there!
→ Welcome to My Live Stream!

See how easily this can work for us in the future? We do not have to repeat this over and over again. Let's make this function a little more interesting by giving it a parameter. Right at the top line, where it says "def say_hi()"? Let us add a parameter here. Type in the word 'name' as a parameter within the parenthesis. Now, the word should be greyed out to confirm that Python has understood the same as a parameter.

Now, you can use this to your advantage and further personalize the greetings to something like this:

If you are doing it on shell you need to delete the previously defined function with

```
Del sayhi
```

Now let's write the new function

```
def say_hi(user):
        print(f"Hello there, {user}!")
        print('Welcome to My Live Stream!')


user = input("Please enter your name to begin: ")
say_hi(user)
```

The output would now ask the user regarding their name. This will then be stored into a variable called user. Since this is a string value, say_hi() should be able to accept this easily. By passing 'user' as an argument, we get this as an output:

Please enter your name to begin: Johnny
→ Hello there, Johnny!
→ Welcome to My Live Stream!

Now that's more like it! Personalized to perfection. We can add as many lines as we want, the function will continue to update itself and provide greetings to various users with different names.

There may be times where you may need more than just the user's first name. You might want to inquire about the last name of the user as well. To add to that, add this to the first line and follow the same accordingly:

```
def say_hi(first_name, last_name):

    print(f"Hello there, {first_name} {last_name}!")

    print('Welcome to My Live Stream!')
```

```
first_name = input("Enter your first name: ")

last_name = input("Enter your last name: ")

say_hi(first_name, last_name)
```

Now, the program will begin by asking the user for their first name, followed by the last name. Once that is sorted, the program will provide a personalized greeting with both the first and last names.

However, these are positional arguments, meaning that each value you input is in order. If you were to change the positions of the names for John Doe, Doe will become the first name and John will become the last name. You may wish to remain a little careful on that.

Hopefully, now you have a good idea of what functions are and how you can access and create them. Now, we will jump towards a more complex front of 'return' statements.

"Wait! There's more?" Well, I could have explained; however, you may not have understood it completely. Since we have covered all the bases, it is appropriate enough for us to see exactly what these are and how these gel along with functions.

Return Statement

Return statements are useful when you wish to create functions whose sole job is to return some values. These could be for users or for programmers alike. It is a lot easier if we do this instead of talk about theories, so let's jump back to our PyCharm and create another function.

Let us start by defining a function called 'cube' which will basically multiply the number by itself three times. However, since we want Python to return a value, we will use the following code:

```python
def cube(number):
    return number * number * number
```

By typing 'return' you are informing Python that you wish for it to return a value to you that can later be stored in a variable or used elsewhere. It is pretty much like the input() function where a user enters something and it gets returned to us.

```python
def cube(number):
    return number * number * number

> number = int(input("Enter the number: "))
> print(cube(number))
```

148

Go ahead and try out the code to see how it works. It is not necessary that we define functions such as these. You can create your own complex functions that convert kilos into pounds, miles into kilometers, or even carry out far greater and more complex jobs. The only limit is your imagination. The more you practice, the more you explore.

With that said, it is time to say goodbye to the world of functions and head into the advanced territories of Python. By now, you already have all you need to know to start writing your own codes.

Multiple Parameters

You can assign two or more parameters in a function. For example:

```
def simpOp(x, y):
    z = x + y


> simpOp(1, 2)
> print(z)
→ 3
```

Lambada function

Anonymous Functions or Lambda

Using an anonymous function is a convenient way to write one-line functions that require arguments and return a value. It uses the keyword lambda. Despite having a purpose of being a one liner, it can have numerous parameters. For example:

```
average = lambda x, y, z: (x + y + z) / 3
x = average(10, 20, 30)
```

X

    → 20.0

average(12, 51, 231)

    → 98.0

Global variables

Global variables utilized in Python are a multipurpose variable used in any part of the Python environment. The variable used can operate in your program or module while in any part.

Local variables

Unlike global variables, local variables are used locally, declared within a Python function or module, and utilized solely in a specific program or Python module. When implemented outside particular modules or tasks, the Python interpreter will fail to recognize the units henceforth throwing an error message for undeclared values.

Python has two built-in methods named globals() and locals(). They allow you to determine whether a variable is either part of the global namespace or the local namespace. The following example shows how to use these methods:

```python
def calc():


    global place
    place = "Rome"
    name = "Leo"
```

```python
        print("place in global:", 'place' in globals())
        print("place in local :", 'place' in locals())
        print("name in global :", 'name' in globals())
        print("name in local  :", 'name' in locals())
    return

place = "Berlin"
print(place)
calc()
```

Chapter 5: Modules in Python

What are the Modules?

In Python, a module is a portion of a program (an extension file) that can be invoked through other programs without having to write them in every program used. Besides, they can define classes and variables. These modules contain related sentences between them and can be used at any time. The use of the modules is based on using a code (program body, functions, variables) already stored on it called import. With the use of the modules, it can be observed that Python allows simplifying the programs a lot because it allows us to simplify the problems into a smaller one to make the code shorter so that programmers do not get lost when looking for something in hundreds of coding lines when making codes.

How to Create a Module?

To create a module in Python, we don't need a lot; it's very simple.

For example: if you want to create a module that prints a city, we write our code in the editor and save it as "mycity.py".

Once this is done, we will know that this will be the name of our module (omitting the .py sentence), which will be assigned to the global variable __city__.

But, beyond that, we can see that the file "mycity.py" is pretty simple and not complicated at all, since the only thing inside is a function called "print_city" which will have a string as a parameter, and what it will do is to print "Hello, welcome to", and this will concatenate with the string that was entered as a parameter.

Locate a Module

When importing a module, the interpreter automatically searches the same module for its current address, if this is not available, Python (or its interpreter) will perform a search on the PYTHONPATH environment variable that is nothing more than a list containing directory names with the same syntax as the environment variable.

If in any particular case, these previous actions failed, Python would look for a default UNIX path (located in /user/local/lib/python on Windows).

The modules are searched in the directory list given by the variable sys.path.

This variable contains the current directory, the PYTHONPATH directory, and the entire directory that comes by default in the installation.

Import Statement

This statement is used to import a module. Through any Python code file, its process is as follows: The Python interpreter searches the file system for the current directory where it is executed. Then, the interpreter searches for its predefined paths in its configuration. When it meets the first match (the name of the module), the interpreter automatically executes it from start to finish. When importing a module for the first time, Python will generate a compiled .pyc extension file. This extension file will be used in the following imports of this module. When the interpreter detects that the module has already been modified since the last time it was generated, it will generate a new module.

You must save the imported file in the same directory where Python is using the import statement so that Python can find it.

As we could see in our example, importing a module allows us to improve the functionalities of our program through external files.

Now, let's see some examples. The first one is a calculator where will create a module that performs all the mathematical functions and another program that runs the calculator itself.

The first thing we do is the module "calculator.py" which is responsible for doing all the necessary operations. Among them are addition, subtraction, division, and multiplication, as you can see.
We included the use of conditional statements such as if, else, and elif. We also included the use of exceptions so that the program will not get stuck every time the user enters an erroneous value at the numbers of the calculator for the division.
After that, we will create a program that will have to import the module previously referred to so that it manages to do all the pertinent mathematical functions.
But at this time, you might be thinking that the only existing modules are the ones that the programmer creates. The answer is no since Python has modules that come integrated into it.
With them, we will make two more programs: the first one is an improvement of the one that we have just done, and the second one will be an alarm that will print on screen a string periodically.

Example

Create a python module called dummymodule.py and write the following inside

```
def testF():
    print("this is a module, goodbuy")
```

save the module in the python installation directory.

In the shell

```
Import dummymodule
```

Now call the function

```
dummymodule.testF()
```

You have used your first module.

Module example One

The first thing that was done was to create the module, but at first sight, we have a surprise, which is that math was imported.

What does that mean to us?
Well, that we are acquiring the properties of the math module that comes by default in Python.

We see that the calculator function is created that has several options.
If the op value is equal to 1, the addition operation is made.
If it is equal to 2, the subtraction operation is made, and so on.

But so new is from op is equal to 5 because, if this is affirmative, then it will return the value of the square root of the values num1 and num2 through the use of math.sqrt(num1), which returns the result of the root.

Then, if op is equal to 6, using functions "math.radians()" which means that num1 or num2 will become radians since that is the type of value accepted by the functions "math.sin()", meaning that the value of the sin of num1 and num2 will return to us, which will be numbers entered by users arbitrarily who will become radians and then the value of the corresponding sin.

The last thing will be to create the main program, as it can be seen next:
Here, we can see the simple program, since it only imports the module "calculator.py", then the variables num1 and num2 are assigned the value by using an input.
Finally, an operation to do is chosen and to finish is called the calculator function of the calculator module to which we will pass three parameters.

Module example Two

We are going to create a module, which has within itself a function that acts as a chronometer in such a way that it returns true in case time ends.
In this module, as you can see, another module is imported, which is called as "time", and as its name refers, functions to operate with times, and has a wide range of functions, from returning dates and times to help to create chronometers, among others.
The first thing we do is to create the cron() function, which starts declaring that the start Alarm variables will be equal to time.time, which means that we are giving an initial value to this function o know the exact moment in which the function was initialized to then enter into an infinite cycle.
Since the restriction is always True, therefore, this cycle will never end, unless the break command is inside it.
Then, within the while cycle, there are several instructions.
The first is that the final variable is equal to time.time() to take into account the specific moment we are located and, therefore to monitor time.
After that, another variable is created called times, and this acquires the value of the final minus start Alarm.

But you will be wondering what the round function does. It rounds up the values; we do that to work easier. But this is not enough, therefore, we use an if since, if the subtraction between the end and the beginning is greater or equal to 60, then one minute was completed, and what happens to this?

Why 60?

This is because the time module works with a second and for a minute to elapse, 60 seconds have to be elapsed, therefore, the subtraction between the end and the beginning has to be greater than or equal to 60, in the affirmative case, True will be returned and finally, we will get out of the infinite cycle.

Once the alarm module is finished, we proceed to make the program, as we can see below:

We can see that the program imports two modules, the one we have created, the alarm and the time module.

The first thing we do is to create the variable s as an input which tells the user if he wants to start.

If the answer is affirmative, then the variable h representing the time will be equal to "time.strftime ("%H:%M:%S")", which means that we are using a function of the time module that returns the hour to use in the specified format so that it can then be printed using the print function.

The next action is to use the alarm module using the command alarm.cron(), which means that the cron() function is being called.

When this function is finished, the time will be assigned to the variable h, again, to finish printing it and being able to observe its correct operation.

As a conclusion of this chapter, we can say that the modules are fundamental for the proper performance of the programmer since they allow to make the code more legible, in addition, that it allows subdividing the problems to attack them from one to one and thus to carry out the tasks easily.

Chapter 6: Files Handling in Python

The next thing that we need to focus on when it comes to working with Python is making sure we know how to work and handle files. It may happen that you are working with some data and you want to store them while ensuring that they are accessible for you to pull up and use when they are needed later. You do have some choices in the way that you save the data, how they are going to be found later on, and how they are going to react in your code.

When working with files, you will find with this one that there are a few operations or methods that you are able to choose when it comes to working with files. And some of these options will include:

- Closing up a file you are working on.
- Creating a brand new file to work on.
- Seeking out or moving a file that you have over to a new location to make it easier to find.

Creating new files

The first task that we are going to look at doing here is working on creating a file. It is hard to do much of the other tasks if we don't first have a file in place to help us out. if you would like to be able to make a new file and then add in some code into it, you first need to make sure the file is opened up inside of your IDLE. Then you can choose the mode that you would like to use when you write out your code.

When it comes to creating files on Python, you will find there are three modes that you are able to work with. The three main modes that we are going to focus on here include append (a), mode(x), and write(w).

Any time that you would like to open up a file and make some changes to it, then you would want to use the write mode. This is the easiest out of the three to work with. The write method is going to make it easier for you to get the right parts of the code set up and working for you in the end.

The write function is going to be easy to use and will ensure that you can make any additions and changes that you would like to the file. You can add in the new information that you would like to the file, change what is there, and so much more. If you would like to see what you can do with this part of the code with the write method, then you will want to open up your compiler and do the following code:

```
#file handling operations

#writing to a new file hello.txt

f = open('hello.txt', 'w', encoding = 'utf-8')

f.write("Hello Python Developers!")

f.write("Welcome to Python World")

f.flush()

f.close()
```

From here, we need to assess what you can do with the directories that we are working with. The default directory is always going to be the current directory. You are able to go through and switch up the directory where the code information is stored, but you have to take the time, in the beginning, to change that information up, or it isn't going to end up in the directory that you would like.

Whatever directory you spent your time in when working on the code is the one you need to make your way back to when you want to find the file later on. If you would like it to show up in a different directory, make sure that you move over to that one before you save it and the code. With the option that we wrote above, when you go to the current directory (or the directory that you chose for this endeavor, then you will be able to open up the file and see the message that you wrote out there.

For this one, we wrote a simple part of the code. You, of course, will be writing out codes that are much more complicated as we go along. And with those codes, there are going to be times when you would like to edit or overwrite some of what is in that file. This is possible to do with Python, and it just needs a small change to the syntax that you are writing out. A good example of what you can do with this one includes:

```
#file handling operations
```

```
#writing to a new file hello.txt
f = open('hello.txt', 'w', encoding = 'utf-8')
f.write("Hello Python Developers!")
f.write("Welcome to Python World")
mylist = ["Apple", "Orange", "Banana"]
#writelines() is used to write multiple lines into the file
f.write(mylist)
f.flush()
f.close()
```

The example above is a good one to use when you want to make a few changes to a file that you worked on before because you just need to add in one new line. This example wouldn't need to use that third line because it just has some simple words, but you can add in anything that you want to the program, just use the syntax above and change it up for what you need.

What are the binary files?

One other thing that we need to focus on for a moment before moving on is the idea of writing out some of your files and your data in the code as a binary file. This may sound a bit confusing, but it is a simple thing that Python will allow you to do. All that you need to do to make this happen is to take the data that you have and change it over to a sound or image file, rather than having it as a text file.

With Python, you can change any of the code that you want into a binary file. It doesn't matter what kind of file it was in the past. But you do need to make sure that you work on the data in the right way to ensure that it is easier to expose in the way that you want later on. The syntax that is going to be needed to ensure that this will work well for you will be below:

```
# write binary data to a file
# writing the file hello.dat write binary mode
F = open('hello.dat', 'wb')
# writing as byte strings
f.write("I am writing data in binary file!/n")
f.write("Let's write another list/n")
f.close()
```

If you take the time to use this code in your files, it is going to help you to make the binary file that you would like. Some programmers find that they like using this method because it helps them to get things in order and will make it easier to pull the information up when you need it.

Opening your file up

So far, we have worked with writing a new file and getting it saved, and working with a binary file as well. In these examples, we got some of the basics of working with files down so that you can make them work for you and you can pull them up any time that you would like.

Now that this part is done, it is time to learn how to open up the file and use it, and later even make changes to it, any time that you would like. Once you open that file up, it is going to be so much easier to use it again and again as much as you would like. When you are ready to see the steps that are needed to open up a file and use it, you will need the following syntax.

```
# read binary data to a file
#writing the file hello.dat write append binary mode
with open("hello.dat", 'rb') as f:
data = f.read()
```

```
text = data.decode('utf-8')

print(text)
```

The output that you would get form putting this into the system would be like the following:

- → Hello, world!
- → This is a demo using with
- → This file contains three lines
- → Hello world
- → This is a demo using with
- → This file contains three lines.
- → Seeking out a file you need

And finally, we need to take a look at how you can seek out some of the files that you need in this kind of coding language. We already looked at how to make the files, how to store them in different manners, how to open them and rewrite them, and then how to seek the file. But there are times where you are able to move one of the files that you have over to a new location.

For example, if you are working on a file and as you do that, you find that things are not showing up the way that you would like it to, and then it is time to fix this up. Maybe you didn't spell the time of the identifier the right way, or the directory is not where you want it to be, then the seek option may be the best way to actually find this lost file and then make the changes, so it is easier to find later on.

With this method, you are going to be able to change up where you place the file, to ensure that it is going to be in the right spot all of the time or even to make it a bit easier for you to find it when you need. You just need to use syntax like what is above to help you make these changes.

Working through all of the different methods that we have talked about here is going to help you to do a lot of different things inside of your code. Whether you would like to make a new file, you want to change up the code, move the file around, and more; you will be able to do it all using the codes that we have gone through here.

Chapter 7: Exception Handling in Python

The next topic that we need to spend some time exploring in this guidebook is the idea of exception handling. There are going to be times when your code tries to show up some errors or other problems inside the code that you are doing, and this is where the exception is going to occur. Knowing when to recognize these exceptions, how to handle them, and even how to make some of your own can make a big difference in how well you are able to do some of your own coding in Python.

A good example of an exception that your compiler is automatically going to raise up is when you or the user is to divide it by zero. The compiler is then going to recognize that this is something the user is not able to do, and it is going to send out that exception as an alert. It can also be something that is going to be called up if you, as the programmer, are trying to call up a function and the name is not spelled in the proper manner so there is no match present to bring up.

There are a few different exceptions that are automatically found in your Python library. It is a good idea to take some time to look through them and recognize these exceptions so you can recognize them later on. Some of the most common exceptions that you need to worry about include:

- Finally—this is the action that you will want to use to perform cleanup actions, whether the exceptions occur or not.
- Assert—this condition is going to trigger the exception inside of the code.
- Raise—the raise command is going to trigger an exception manually inside of the code.
- Try/except—this is when you want to try out a block of code and then it is recovered thanks to the exceptions that either you or the Python code rose.

Raising an Exception

The first thing that we need to take a look at, now that we know a bit more about these exceptions and what they mean, is how to write one out, and some of the steps that you can use if one of these does end up in your own code. If you are going through some of the code writing, and you start to notice that an exception will be raised, know that often the solution is going to be a simple one. But as the programmer, you need to take the time to get this fixed. To help us get started here, let's take a look at what the syntax of the code for raising an exception is all about.

```
> x = 10

> y = 10

> result = x/y #trying to divide by zero

> print(result)
```

The output that you are going to get when you try to get the interpreter to go through this code would be:

```
→ Traceback (most recent call last):
→ File "D: \Python34\tt.py", line 3, in <module>
→ result = x/y
→ ZeroDivisionError: division by zero
```

As we take a moment to look at the example that we have here, we can see that the program is going to bring up an exception for us, mainly because we are trying to divide a number by zero and this is something that is not allowed in the Python code (and any other coding language for that matter). If you decide not to make any changes at this point, and you go ahead and run the program as it is, you could end up with the compiler sending you an error message. The code is going to tell the user the problem, but as you can see, the problem is not listed out in an easy-to-understand method and it is likely the user will have no idea what is going on or how they can fix the problem at all.

With that example that we worked on above, you have some options. You can choose to leave the message that is kind of confusing if you don't know any coding, or you can add in a new message that is easier to read and explains why this error has been raised in the first place. It won't have a lot of numbers and random letters that only make sense to someone who has done coding for a bit, which makes the whole thing a bit more user-friendly overall. The syntax that you can use to control the message that your user is going to see includes:

```
> x = 10
> y = 0
> result = 0
> try:
> result = x/y
> print(result)
→ except ZeroDivisionError:
> print("You are trying to divide by zero.")
```

Take a look at the two codes above. The one that we just did looks a little bit similar to the one above it, but this one has a message inside. This message is going to show up when the user raises this particular exception. You won't get the string of letters and numbers that don't make sense, and with this one, the user will know exactly what has gone wrong and can fix that error.

Can I define my own exceptions?

In the examples above, we took some time to define and handle the exceptions that the compiler offered to us and are already found in the Python library. Now it is time for us to take it a bit further and learn how to raise a few of our own exceptions in any kind of code that we want to write. Maybe you are working on a code that only allows for a few choices to the user, one that only allows them to pick certain numbers or one that only allows them to have so many chances at guessing. These are common things that we see when we work with gaming programs but can work well in other programs that you design.

When you make these kinds of exceptions, the compiler is going to have to be told that an exception is being raised, because it is not going to see that there is anything wrong in this part of the code. The programmer has to go in and let the compiler know what rules it has to follow, and what exceptions need to be raised in the process. A good example of the syntax that you can use to make this happen in your own code will be below:

```
class CustomException(Exception):
    def init_(self, value):
        self.parameter = value

    def str_(self):
    return repr(self.parameter)

    CustomException("This is a CustomError!")
        except CustomException as ex:
        print("Caught:", ex.parameter)
```

In this code, you have been successful in setting up your own exceptions and whenever the user raises one of these exceptions, the message of "Caught: This is a CustomError!" is going to come up on the screen. This is the best way to show your users that you have added a customer exception into the program, especially if this is just one that you personally created for this part of the code, and not one that the compiler is going to recognize on its own.

Just like with the other examples that we went through, we worked with some generic wording just to show how exceptions are able to work. You can easily go through and change this up so that you get a message that is unique for the code that you are writing and will explain to the user what is going on when they get the error message to show up.

Learning how to work with some of the exceptions that can come up in your code is one of the best ways to make sure that your codes work the way that you want, that the user is going to like working with your program, and that everything is going to proceed as normal and do what you want. Take some time to practice these examples and see how they can work for you in order to handle any of the exceptions that come up in your code.

Chapter 8: Objects and Classes in Python

We have already mentioned that Python is an object-oriented programming language. There are other languages that are procedure-oriented that emphasize functions, but in Python, the stress is on objects. But then, what is an object?

Simply put, an object is a collection of methods (functions) that act on data (variables) which are also objects. The blueprint for these objects is a class.

Consider a class a sketch or a prototype that has all the details about an object. If your program were a car, a class would contain all the details about the design, the chassis, where tires are, and what the windshield is made of. It would be impossible to build a car without a class defining it. The car is the object.

Because many cars can be built based on the prototype, we can create many objects from a class. We can also call an object an instance of a class, and the process by which it is created is called instantiation.

Defining a Class

Classes are defined using the keyword class. Just like a function, a class should have a documentation string (docstring) that briefly explains what the class is and what it does. While the docstring is not mandatory, it is a good practice to have it. Here is a simple definition of a class called NewClass:

```
class NewClass:
    """This is the docstring of the class NewClass that we
    Just created. Our program now has a new class"""
    pass
```

When you create a new class, a new local namespace that defines all its attributes is created. Attributes in this case may include functions and data structures. In it, it will contain special attributes that start with __ (double underscores) e.g. __doc__ that defines the docstring of the class.

When a class is defined, a new class object with the same name is created. The new class object is what we can use to access the different attributes and to instantiate the new objects of our new class.

Creating a new class

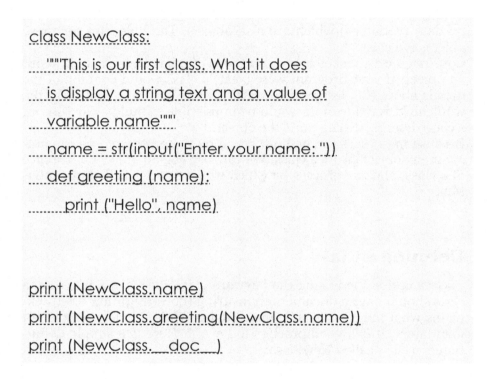

```
class NewClass:
    """This is our first class. What it does
    is display a string text and a value of
    variable name"""
    name = str(input("Enter your name: "))
    def greeting (name):
        print ("Hello", name)


print (NewClass.name)
print (NewClass.greeting(NewClass.name))
print (NewClass.__doc__)
```

What does your console display when you run the script?

Creating an Object

So far, we have learned that we can access the different attributes of a class using the class objects. We can use these objects to also instantiate new instances of that class using a procedure a lot similar to calling a function.

→ MyObject = NewClass()

In the example above, a new instance object called MyObject is created. This object can be used to access the attributes of the class NewClass using the class name as a prefix.

The attributes in this case may include methods and variables. The methods of an object are the corresponding functions of a class meaning that any class attribute function object defines the methods for objects in that class.

For instance, because NewClass.greeting is a function object and an attribute of NewClass, MyObject.greeting will be a method object.

Creating an Object

```
class NewClass:
    """This is our first class. What it does
    is display a string text and a value of
    variable name"""
    name = str(input("Enter your name: "))
    def greeting (name):
        print ("Hello", name)




MyObject = NewClass() #Creates a new NewClass object
print (NewClass.greeting)
print (MyObject.greeting)
MyObject.greeting() # Calling function greeting()
```

the name parameter is within the function definition of the class, but we called the method using the statement MyObject.greeting() without specifying any arguments and it still worked. This is because when an object calls a method defined within it, the object itself passes as the first argument. Therefore, in this case, MyObject.greeting() translates to NewClass.greeting(MyObject).

Generally speaking, when you call a method with a list of x arguments, it is the same as calling the corresponding function using an argument list created when the method's object is inserted before the first argument.

As a result, the first function argument in a class needs to be the object itself. In Python, this is typically called self but it can be assigned any other name. It is important to understand class objects, instance objects, function objects, and method objects and what sets them apart.

Constructors

In python, the __init__() function is special because it is called when a new object of its class is instantiated. This object is also called a constructor because it is used to initialize all variables.

Constructors

MyObject.greeting() # class ComplexNumbers:

 def __init__(self, x = 0, y = 0):

  self.real = x

  self.imagined = y

 def getNumbers(self):

  print ("Complex numbers are: {0}+{1}j".format(self.real, self.imagined))

- → Object1 = ComplexNumbers(2, 3) #Creates a new ComplexNumbers object
- → Object1.getNumbers() #Calls getNumbers() function
- → Object2 = ComplexNumbers(10) #Creates another ComplexNumbers object
- → Object2.attr = 20 #Creates a new attribute 'attr'
- → print ((Object2.real, Object2.imagined, Object2.attr))
- → Object1.attr #Generates an error because c1 object doesn't have attribute 'attr'

In the above exercise, we have defined a new class that represents complex numbers. We have defined two functions, the __init__() function that initializes the variables and the getNumbers() function to properly display the numbers.

Note that the attributes of the objects in the exercise are created on the fly. For instance, the new attribute attr for Object2 was created but one for Object1 was not (hence the error).

Deleting Attributes and Objects

You can delete the attributes of an object or even the object itself at any time using the statement del.

Deleting Attributes and Objects

```
class ComplexNumbers:

    def __init__(self, x = 0, y = 0):

        self.real = x

        self.imagined = y

    def getNumbers(self):

        print ("Complex numbers are: {0}+{1}j".format(self.real, self.imagined))

Object1 = ComplexNumbers(2, 3) #Creates a new ComplexNumbers object

Object1.getNumbers() #Calls getNumbers() function

Object2 = ComplexNumbers(10) #Creates another ComplexNumbers object

Object2.attr = 20 #Creates a new attribute 'attr'

print ((Object2.real, Object2.imagined, Object2.attr))

del ComplexNumbers.getNumbers

Object1.getNumbers()
```

The error you get when you run the script shows that the attribute getNumbers() has been deleted. Note, however, that since a new instance is created in memory when a new instance of the object is created, the object may continue to exist in memory even after it is deleted until the garbage collector automatically destroys unreferenced objects.

Chapter 9: Inheritance and Polymorphism

In Python, a class is defined using the keyword Class, the same way a function is defined with the keyword def. Therefore, if we were to define a class called ClassName, our syntax would look like this:

```
class ClassName:
    "Class Documentation String"
    Class_Suite
```

In this syntax, the keyword class is used to define a new class, followed by the class name ClassName, and a colon. The class documentation string is essentially a definition or description of the class. The Class_Suite is representative of the entirety of the defining class members, component statements, functions, and data attributes of the class.

Since a class is used to create objects, it is best seen as a way to add consistency to programs created in Python so that they are cleaner, more efficient, and most importantly, functional. To create a class that can be instantiated anywhere in your code, it must be defined at a module's top-level.

Creating a Class in Python

Now that we are familiar with the syntax for creating a class in Python, we'll introduce a class example called Dog.

```
class Dog:
"Dog class"
    var1 = "Bark"
    var2 = "Jump"
```

Class Declaration and Definition

In Python 3, there is no difference between class declaration and class definition. This is because the two occur simultaneously. Class definition follows declaration and documentation string as demonstrated in our example.

Class Methods and Attributes

A class, defined and created, is not complete unless it has some functionality. Functionalities in a class are defined by setting their attributes, which are best seen as containers for functions and data related to those attributes.

Class attributes include data members like variables and instance variables and methods found using the dot notation. Here are their definitions:

• Class variable: This is a variable shared by all the class instances and objects.

• Instance variable: This is a variable unique to an instance of a class. It is typically defined within a method and will only be applicable to the instance of that class.

• Method: Also called a function, a method is defined in a class and defines the behavior of an object.

Class Attributes

A class attribute is a functional element that belongs to another object and is accessed via dotted-attribute notation. In Python, complex numbers have data attributes while lists and dictionaries have functional attributes. When you access an attribute, you can also access an object that may have attributes of its own.

Class attributes are linked to the classes in which they are defined. The most commonly used objects in OOP are instance objects. Instance data attributes are the primary data attributes that are used. You'll find most use for class data attributes when you require a data type, which does not need any instances.

Class Data Attributes

Data attributes are the class variables that are defined by the programmer. They can be used like any other variable when creating the class. Methods can update them within the class. Programmers

know these types of attributes better as static members, class variables, or static data. They represent data tied to the class object they belong to and are independent of any class instances.
Example of using class data attributes (xyz):

```
class ABC:
    xyz = 10
    print(ABC.xyz)
    ABC.xyz = ABC.xyz + 1
    print (ABC.xyz)
```

The output of this example code will be 11.

Python Class Inheritance

Inheritance is a feature in object-oriented programming that allows a class to inherit methods and attributes from a parent class, also referred to as base class. This is a very handy feature in programming as it allows the programmer to create a suite of functionality for a single class and then pass them on to sub-classes or child classes. As a result, the program will be able to create new and even overwrite existing functionalities in a child class without affecting the functionality of the parent class.

The sub-classes that inheritance creates will feature the specializations of the parent classes. There are four types of inheritances in Python: single, multilevel, hierarchical, and multiple inheritances.

Single Inheritance

Programming classes that have no inheritance features can be accurately referred to as object-based programming. The program, when run, should create new abstract data types, each with its own operations. However, what separates object-oriented programming from object-based programming is inheritance. Single inheritance is when a class or subclass inherits methods and attributes from one parent class.

Multiple Inheritance

In multiple inheritance, a child class or subclass inherits methods and attributes from multiple classes. For instance, a class C can inherit the features of both class A and B, in the same way that a child inherits the characteristics of both the mother and father. In some cases, a child class can inherit the features and functionalities of more than two-parent classes.

There is no limit to the number of parent classes from which a child class can inherit methods and attributes. Note that while multiple inheritances is best known to reduce program redundancy, it may also introduce a higher level of complexity and ambiguity to the program and must be properly thought-out during program design before implementation.

Multilevel Inheritance

We have already established that it is possible for a class in Python to inherit features of multiple parent classes. When a class inherits the methods and functions of other classes that also inherit them from other classes, the process is known as multilevel inheritance. Like in C++ and other object-oriented programming languages, Python allows for multilevel inheritance implemented at any depth.

Hierarchical Inheritance

A hierarchical inheritance occurs when more than one class is derived from a single parent or base class. The features inherited by the sub-class or the child class are included in the parent class. What sets hierarchical inheritance apart from a multi-level inheritance is the order in which the relationship between the classes is established. In multilevel inheritance, the order can be haphazard and parent classes can inherit features from previous child classes.

Why is Inheritance Useful in Python Programming?

Inheritance is a very handy feature of object-oriented programming because it allows a programmer to easily adhere to one of software development's most important– Don't Repeat Yourself (DRY). Simply put, implementing class inheritance in your programs is the most efficient way to get more done with fewer lines of code and less repetition.

Inheritance will also compel you to pay closer attention to the design phase of programming to ensure that you write a program code that is clean, minimalist, and effective.

Another use of inheritance is adding functionality to various sections of your program.

Inheritance Example

In Python, this is done by deriving classes. Let's say we have class called SportsCar.

```
class Vehicle(object):
        def __init__(self, makeAndModel, prodYear, air-
Conditioning):
                self.makeAndModel = makeAndModel
                self.prodYear = prodYear
                self.airConditioning = airConditioning
                self.doors = 4
        def honk(self):
                print "%s says: Honk! Honk!" %
self.makeAndModel
```

Now, below that, create a new class called SportsCar, but instead of deriving
object, we're going to derive from Vehicle.

```
class SportsCar(Vehicle)
```

```
def __init__(self, makeAndModel, prodYear, air-
Conditioning):
        self.makeAndModel = makeAndModel
        self.prodYear = prodYear
        self.airConditioning = airConditioning
        self.doors = 4
```

Leave out the honk function, we only need the constructor function here.
Now declare a sports car. I'm just going to go with the Ferrari.

```
ferrari = SportsCar("Ferrari Laferrari", 2016, True)
```

Now test this by calling

```
ferrari.honk()
```

and then saving and running. It should go off without a hitch. Why is this? This is because the notion of inheritance says that a child class derives functions and class variables from a parent class. Easy enough concept to grasp. The next one is a little tougher.

Class Polymorphism and Abstraction

In Computer Science, polymorphism and abstraction are advanced programming features that extend the application and usefulness of inheritance.
Polymorphism means that should a class Y inherit from class X, it does not necessarily have to inherit everything from that class. It can implement some of the inherited methods and attributes differently. Python, being implicitly polymorphic, can overload operators to grant them the desired behavior based on individual contexts.

The idea of polymorphism is that the same process can be performed in different ways depending upon the needs of the situation. This can be done in two different ways in Python: *method overloading* and *method overriding*.

Method overloading is defining the same function twice with different arguments. For example, we could give two different initializer functions to our Vehicle class. Right now, it just assumes a vehicle has 4 doors. If wewanted to specifically say how many doors a car had, we could make a new initializer function below our current one with an added *doors* argument, like so (the newer one is on the bottom):

def __init__(self, makeAndModel, prodYear, airConditioning):

self.makeAndModel = makeAndModel

self.prodYear = prodYear

self.airConditioning = airConditioning

self.doors = 4

def __init__(self, makeAndModel, prodYear, airConditioning, doors):

self.makeAndModel = makeAndModel

self.prodYear = prodYear

self.airConditioning = airConditioning

self.doors = doors

Somebody now when creating an instance of the Vehicle class can *choose* whether they define the number of doors or not. If they don't, the number of doors is assumed to be 4. *Method overriding* is when a child class *overrides* a parent class's function with its code. To illustrate, create another class which extends Vehicle called Moped. Set the doors to 0, because that's absurd, and set air conditioning to false. The only relevant arguments are make/model and production year. It should look like this:

```
class Moped(Vehicle):
def __init__(self, makeAndModel, prodYear):
self.makeAndModel = makeAndModel
self.prodYear = prodYear
self.airConditioning = False
self.doors = 0
```

Now, if we made an instance of the Moped class and called the honk() method, it would honk. But it is common knowledge that mopeds don't honk, they beep. So let's override the parent class's honk method with our own. This is super simple. We just redefine the function in the child class:

```
def honk(self):
print "%s says: Beep! Beep!" % self.makeAndModel
```

I'm part of the 299,000,000 Americans who couldn't name a make and model of moped if their life depended on it, but you can test out if this works for yourself but declaring an instance of the Moped class and trying it out."

Abstraction

Abstraction is generally a net positive for a large number of applications that are being written today, and there's a reason Python and other object-oriented programming languages are incredibly popular. Abstraction is the process of simplifying complex realities by modeling classes to handle specific problems. An abstract class cannot be instantiated and you can neither create class instances nor objects for them. Abstract classes are designed to inherit all or only specific features from a base class. Abstraction innately makes the language easier to understand, read, and learn. Though it makes the language a tad bit less powerful by taking away some of the power that the user has over the entire computer architecture, this is traded instead for the ability to program quickly and efficiently in the language, not wasting time dealing with trivialities like memory addresses or things of the like. These apply in Python because, well, it's incredibly simple. You can't get down into the nitty-gritty of the computer, or do much with memory allocation or even specifically allocate an array size too easily, but this is a tradeoff for amazing readability, a highly secure language in a highly secure environment, and ease of use with programming. Compare the following:

```
snippet of code from C:
#include <stdio.h>
int main(void) {
printf("hello world");
return 0;
}
```

to the Python code for doing the same:

```
print ("hello world")
# That's it. That's all there is to it.
```

Encapsulation

The last major concept in object-oriented programming is that of encapsulation. This one's the easiest to explain. This is the notion that common data should be put together, and that code should be modular. I'm not going to spend long explaining this because it's a super simple concept. The entire notion of classes is as concise of an example as you can get for encapsulation: common traits and methods are bonded together under one cohesive structure, making it super easy to create things of the sort without having to create a ton of super-specific variables for every instance. Well, there we go.

Conclusion

Programming isn't just about getting a PC to get things done. It is tied in with composing code that is helpful to people. Great programming is saddling complexity by composing code that rhymes with our instincts. Great code will be code that we can use with a negligible amount of setting.

The most important thing for you to do is to practice programming in Python. If you have read until here then you have already absorbed quite much. You need to practice all the things you have learned to make sure you consolidate that knowledge (i.e. make it stick).

Knowledge is useless without application. Learning how to program without actual programming will only waste the time you invested here. It is like learning how to ride a bike by reading books or articles about it - that will never be enough! You need to ride a bike to learn how to ride a bike.

Also, make sure to familiarize yourself with useful resources you can easily refer to when you need help. There are three obvious ones: Python's documentation, Stack Exchange.

During your programming journey, you will encounter seemingly impossible problems. Always get help if you encounter those problems. And during those times, never hesitate to reach out for help.

- PYTHON FOR DATA ANALYSIS & ANALYTICS -

by

TechExp Academy

2020 © Copyright

Introduction

Data analysis plays an important role in many aspects of life today. From the moment you wake up, you interact with data at different levels. A lot of important decisions are made based on data analytics. Companies need data to help them meet many of their goals. As the population of the world keeps growing, its customer base keeps expanding. In light of this, they must find ways of keeping their customers happy while at the same time meeting their business goals.

Given the nature of competition in the business world, it is not easy to keep customers happy. Competitors keep preying on each other's customers, and those who win have another challenge ahead - how to maintain the customers lest they slide back to their former business partners. This is one area where data analysis comes in handy.

To understand their customers better, companies rely on data. They collect all manner of data at each point of interaction with their customers. Data are useful in several ways. The companies learn more about their customers, thereafter clustering them according to their specific needs. Through such segmentation, the company can attend to the customers' needs better and hope to keep them satisfied for longer.

But, data analytics is not just about customers and the profit motive. It is also about governance. Governments are the biggest data consumers all over the world. They collect data about citizens, businesses, and every other entity that they interact with at any given point. This is important information because it helps in a lot of instances.

For planning purposes, governments need accurate data on their population so that funds can be allocated accordingly. Equitable distribution of resources is something that cannot be achieved without proper data analysis. Other than

planning, there is also the security angle. To protect the country, the government must maintain different databases for different reasons. There are high profile individuals who must be accorded special security detail, top threats who must be monitored at all times, and so forth. To meet the security objective, the government has to obtain and maintain updated data on persons of interest at all times.

There is so much more to data analysis than the corporate and government decisions. As a programmer, you are venturing into an industry that is challenging and exciting at the same time. Data doesn't lie unless it is manipulated, in which case you need insane data analysis and handling skills. As a data analyst, you will come across many challenges and problems that need solutions that can only be handled through data analysis. The way you interact with data can make a big difference, bigger than you can imagine.

There are several tools you can use for data analysis. Many people use Microsoft Excel for data analysis and it works well for them. However, there are limitations of using Excel which you can overcome through Python. Learning Python is a good initiative, given that it is one of the easiest programming languages. It is a high-level programming language because its syntax is so close to the nor-mal language we use. This makes it easier for you to master Python concepts.

For expert programmers, you have gone beyond learning about the basics of Python and graduated into using Python to solve real-world problems. Many problems can be solved through data analysis. The first challenge is usually understanding the issue at hand, then working on a data solution for it.

This book follows a series of elaborate books that introduced you to data analysis using Python. Some important concepts have been re-iterated since the beginning of the series to help you remember the fundamentals. Knowledge of Python libraries is indeed important. It is by understanding

these libraries that you can go on to become an expert data analyst with Python.

As you interact with data, you do understand the importance of cleaning data to ensure the outcome of your analysis is not flawed. You will learn how to go about this and build on that to make sure your work is perfect. Another challenge that many organizations have is protecting the integrity of data. You should try and protect your organization from using contaminated data. There are procedures you can put in place to make sure that you use clean data all the time.

We live in a world where data are at the center of many things we do. Data are produced and stored in large amounts daily from automated systems. Learning data analysis through Python should help you process and extract information from data and make meaningful conclusions from them. One area where these skills will come in handy is forecasting. Through data analysis, you can create predictive models that should help your organization meet its objectives.

A good predictive model is only as good as the quality of data introduced into it, the data modeling methods, and more importantly, the dataset used for the analysis. Beyond data handling and processing, one other important aspect of data analysis is visualization. Visualization is about presentation. Your data model should be good enough for an audience to read and understand it at the first point of contact. Apart from the audience, you should also learn how to plot data on different visualizations to help you get a rough idea of the nature of the data you are working with.

When you are done with data analysis, you should have a data model complete with visual concepts that will help in predicting outcomes and responses before you can proceed to the testing phase. Data analysis is a study that is currently in high demand in different fields. Knowing what to

do, as well as when and how to handle data, is an important skill that you should not take for granted. Through this, you can build and test a hypothesis and go on to understand systems better.

Table of Contents

Chapter 1: Data Analysis, the Basics

What is Data Analysis?

Data analysis is defined as a process of cleaning, transforming, and modeling data to discover useful information for business decision-making. The purpose of Data Analysis is to extract useful information from data and taking the decision based upon the data analysis. Whenever we take any decision in our day-to-day life is by thinking about what happened last time or what will happen by choosing that particular decision. This is nothing but analyzing our past or future and making decisions based on it. For that, we gather memories of our past or dreams of our future. So that is nothing but data analysis. Now same thing analyst does for business purposes, is called Data Analysis.

Data analysis can also be thought as a practice where a company can take their raw data and then order and organize it. When the data is organized and run through a predictive model, it is going to help the company extract useful information out of it. The process of organizing and thinking about our data is going to be very important as it is the key to helping us understand what the data does and does not contain at any given time.

Many companies have been collecting data for a long time. They may gather this data from their customers, from surveys, from social media, and many other locations. And while collecting the data is an important step that we need to focus on as well, another thing to consider is what we can do with the data. You can collect all of the data that you would like, but if it just sits in your cloud or a data warehouse and is never mined or used, then it is going to become worthless to you, and you wasted a lot of time and money trying to figure it all out.

This is where data analysis will come in. it is able to take all of that raw data and actually, put it to some good use. It will use various models and algorithms, usually with the help of machine learning and Python, in order to help us to understand what important insights and information are found in our data, and how we are able to utilize these for our own benefit.

Why Data Analysis?

The reason that data analysis and analytics is so important is that it can help a business to optimize their performances overall.

When a company is able to implement good data analysis into their business model, it means that they are able to reduce the costs that they experience on a day to day basis. This happens because the analysis will help them to identify the best, and the most efficient, ways of doing business, and because they are able to store up large amounts of data to help them get all of this process done in a timely manner.

Another benefit of using this data analysis, and why it really does matter for a lot of companies, is that the company can use this process in order to make the best business decisions. These business decisions no longer need to rely on what other companies are doing or on the intuition of key decision-makers. Instead, they rely on the facts and insights provided in the collected data.

Many companies also like to work with the process of data analytics because it will help them learn more about and serve their customers better. Data analytics can help us to analyze customer trends and the satisfaction levels of our customers, which can help the company come up with new, and better, services and products to offer.

Data Analysis Tools

Here are the best tools you can use for data analysis:
- Python
- Xplenty
- IDEA
- Microsoft HDInsight
- Skytree
- Talend
- Splice Machine
- Spark
- Plotly
- Apache SAMOA
- Lumify

Types of Data Analysis: Techniques and Methods

There are different type of Data Analysis, below is a list of the most common ones

- Text Analysis
- Statistical Analysis
- Descriptive Analysis
- Diagnostic Analysis
- Predictive Analysis

Let's look at the them more deeply.

Text Analysis

This is going to be a form of predictive analytics that can help us extract the sentiments behind the text, based on the intensity of the presses on the keys, and the typing speeds.

Statistical Analysis

Statistical analysis involves many aspects of statistics. This analysis uses the concepts in statistics to make certain assumptions and predictions of data that has been collected by a company over a decade. For instance, let us assume that you are a clinical research company, and you have been trying to gather information on certain products being sold by your company. You will conduct thorough research on the type of product, the market you are targeting, and your competitors. Let us assume that you are looking to introduce a painkiller in the market. At such a time, you will try to understand the market better and see what it is that you can offer as compared to the other companies in the market. For you to do this, you will need to conduct a thorough statistical analysis

Descriptive Analysis

This is a process that is used in data mining and business intelligence. In the descriptive analysis, you will be able to look at data and also analyze past events. You will be able to obtain insight into approaching the events that may occur in the future. You will use this process to analyze and mine through the data to determine the reasons why the business has either succeeded or failed in the past. Every department in the company would always use this type of analysis to understand their successes and failures.

Diagnostic Analysis

As the name suggests, this type of analysis is done to "diagnose" or understand why a certain event unfolded and how that event can be prevented from occurring in the future or replicated if needed. For example, web marketing strategies and campaigns often employ social media platforms to get publicity and increase their goodwill. Not all campaigns are as successful as expected; therefore, learning from failed campaigns is just as important, if not more. Companies can run diagnostic analysis on their campaign by collecting data pertaining to the "social media mentions" of the campaign, number of campaign page views, the average amount of time spent on the campaign page by an individual, number of social media fans and followers of the campaign, online reviews and other related metrics to understand why the campaign failed and how future campaigns can be made more effective.

Predictive Analysis

The predictive analysis gets driven by predictive modeling. It isn't exactly a process but more of an approach. In the case of data science, machine learning and predictive analysis go hand in hand. The predictive models normally include machine learning algorithms. These models are trained over a period of time for responding to newer values or data and deliver necessary information for business requirements. Predictive modeling basically overlaps with machine learning.

You can find two kinds of predictive models and they are classification models and regression models which predict class membership and numbers respectively. These models are made up by using algorithms. These algorithms perform statistical analysis and data mining to decide the patterns and trends emerging out of the available data. There are software solutions available for predictive analysis which comes with built-in algorithms that are utilized for making predictive models. These algorithms are called classifiers and they identify what set of categories this data belongs to.

Prescriptive Analysis

Prescriptive analysis works towards optimizing and simulating the data and creating a model for the future. This type of analysis helps in synthesizing big data and will also help you understand the rules of business to help you make predictions about the future.

Data Analysis Process

When it comes to working with data analysis it is a good practice to follow a well defined process. This will ensure that you can handle the data in the proper manner and that it will work the way that we want it to. These are going to include some of the initial phases of cleaning our data, working with whether the data is high enough quality, quality measurement analysis, and then we enter into the main data analysis. Below are shown the main macro phases of a data analysis process:

- Data Requirement Gathering
- Data Collection
- Data Cleaning
- Data Analysis
- Data Interpretation
- Data Visualization

1 - Data Requirement Gathering

It is important to understand what information or data needs to be gathered to meet the business objective and goals. Data organization is also very critical for efficient and accurate data analysis. Some of the categories in which the data can be organized are gender, age, demographics, location, ethnicity, and income. A decision must also be made on the required data types (qualitative and quantitative) and data values (can be numerical or alphanumerical) to be used for the analysis.

2 - Data Collection

Searching for the data that we want to work with. Once we have a good idea of the information that we need, and the business problem that we would like to solve, it is time for us to go through and look for the data. There are a number of places where we are able to find this data, such as in surveys, social media, and more, so going out and searching for it here is going to be the best way to gather it up and have it ready to work with on the later steps.

3 - Data Cleaning

The next step that we need to focus on here is data cleaning. While it may not be as much fun as we see with the algorithms and more that come with data analysis, they are still important. This is the part of the process where we match up records, check for multiples and duplicates in the data, and get rid of anything that is false or does not match up with what we are looking for at this time.

4 - Data Analysis

Once the whole process of making sure you clean the data, and we have done the quality analysis and the measurement, it is time to dive into the analysis that we want to use. There are a ton of different analysis that we can do on the information, and it often will depend on what your goals are in this whole process. We can go through and do some graphical techniques that include scattering plots. We can work with some frequency counts to see what percentages and numbers are present. We can do some continuous variables or even the computation of new variables.

5 - Data Interpretation

Look over the insights and hidden patterns that were found in the data. The whole point of working with data analysis is to make sure that we can take a large amount of data and see what important insights are found in that information. The more that we study the data, and the better the algorithm we choose to work with, the easier it is to find the insights and the hidden values that are inside of it. We can then use this to help us make better and more informed decisions.

6 - Data Visualization

This is not a step that you should miss out on at all. These visuals are going to make it easier for those who are in charge of looking over the information to really see the connections and the relationships that show up in that data. This makes it easier for you to really figure out what the data is saying, and to figure out what decisions you should make based on that.

Chapter 2: Applications of Data Analysis and Analytics

eCommerce

Over 2.6 billion and counting active social media users include customers and potential customers for every company out there. The race is on to create more effective marketing and social media strategies, powered by machine learning, aimed at providing enhanced customer experience to turn prospective customers into raving fans. The process of sifting through and analyzing a massive amount of data has not only become feasible, but it's easy now. The ability to bridge the gap between execution and big data analysis has been supplemented by artificial intelligence marketing solutions. Artificial Intelligence (AI) marketing can be defined as a method of you using artificial intelligence consonants like machine learning on available customer data to anticipate customer's needs and expectations while significantly improving the customer's journey. Marketers can boost their campaign performance and return on investment read a little to no extra effort in the light of big data insights provided by artificial intelligence marketing solutions. The key elements that make AI marketing as powerful are:

• Big data - A marketing company's ability to aggregate and segment a huge dump of data with minimal manual work is referred to as Big Data. The marketer can then leverage the desired medium to ensure the appropriate message is being delivered to the target audience at the right time.

• Machine learning - Machine learning platforms enable marketers to identify trends or common occurrences and gather effective insights and responses, thereby deciphering the root cause and probability of recurring events.

• Intuitive platform – Super fast and easy to operate applications are integral to AI marketing. Artificial intelligence technology is capable of interpreting emotions and communicating like a human, allowing AI-based platforms to understand open form content like email responses and social media.

Predictive Analysis

All artificial intelligence technology-based solutions are capable of extracting information from data assets to predict future trends. AI technology has made it possible to model trends that could previously be determined only retroactively. These predictive analysis models can be reliably used in decision-making and to analyze customers' purchase behavior. The model can successfully determine when the consumer is more likely to purchase something new or reorder an old purchase. The marketing companies are now able to reverse engineer customer's experiences and actions to create more lucrative marketing strategies. For example, FedEx and Sprint are using predictive analytics to identify customers who are at potential risk of deflecting to the competitor.

Smart searches

Only a decade ago, if you type in "women's flip flops" on Nike.com, the probability of you finding what you were looking for would be next to zero. But today's search engines are not only accurate but also much faster. This upgrade has largely been brought on by innovations like "semantic search" and "natural language processing" that enable search engines to identify links between products and provide relevant search results, recommend similar items, and autocorrect typing errors. Artificial intelligence technology and big data solutions can rapidly analyze user search patterns and identify key areas that marketing companies should focus on. In 2015, Google introduced the first Artificial Intelligence-based search algorithm called "RankBrain." Following Google's lead, other major e-commerce websites, including Amazon has incorporated big data analysis and artificial intelligence into their search engines to offer smart search experience for their customers, who can find desired products even when they don't know exactly what they're looking for. Even small e-commerce stores have access to Smart search technologies like "Elasticsearch." The data-as-a-service companies like "Indix" allow companies to learn from other larger data sources to train their product search models.

Recommendation Engines

Recommendation engines have quickly evolved into fan favorites and are loved by the customers just as much as the marketing companies. "Apple Music" already knows your taste in music better than your partner, and Amazon always presents you with a list of products that you might be interested in buying. This kind of discovery aide that can sift through millions of available options and hone in on an individual's needs are proving indispensable for large companies with huge physical and digital inventories.

In 1998, Swedish computational linguist, Jussi Karlgren, explored the practice of clustering customer behaviors to predict future behaviors in his report titled "Digital bookshelves." The same here, Amazon implemented collaborative filtering to generate recommendations for their customers. The gathering and analysis of consumer data paired with individual profile information and demographics, by the predictive analysis based systems allow the system to continually learn and adapt based on consumer activities such as likes and dislikes on the products in real-time. For example, the company "Sky" has implemented a predictive analysis based model that is capable of recommending content according to the viewer's mode. The smart customer is looking for such an enhanced experience not only from their Music and on-demand entertainment suppliers but also from all other e-commerce websites.

Product Categorization and Pricing

E-commerce businesses and marketing companies have increasingly adopted artificial intelligence in their process of categorization and tagging of the inventory. The Marketing companies are required to deal with awful data just as much, if not more than amazingly organized, clean data. This bag of positive and negative examples serves as training resources for predictive analysis based classification tools. For example, different detailers can have different descriptions for the same product, such as sneakers, basketball shoes, trainers, or Jordan's, but the AI algorithm can identify that these are all the same products and tag them accordingly. Or if the data set is missing the primary keyword like skirts or shirts, the artificial intelligence algorithm can identify and classify the item or product as skirts or shirts based solely on the surrounding context.

We are familiar with the seasonal rate changes of the hotel rooms, but with the advent of artificial intelligence, product prices can be optimized to meet the demand with a whole new level of precision. The machine learning algorithms are being used for dynamic pricing by analyzing customer's data patterns and making near accurate predictions of what they are willing to pay for that particular product as well as their receptiveness to special offers. This empowers businesses to target their consumers with high precision and calculated whether or not a discount is needed to confirm the sale. Dynamic pricing also allows businesses to compare their product pricing with the market leaders and competitors and adjust their prices accordingly to pull in the sale. For example, "Airbnb" has developed its dynamic pricing system, which provides 'Price Tips' to the property owners to help them determine the best possible listing price for their property. The system takes into account a variety of influencing factors such as geographical location, local events, property pictures, property reviews, listing features, and most importantly, the booking timings and the market demand. The final decision of the property owner should follow or ignore the provided 'price tips' and the success of the listing are also monitored by the system, which will then process the results and adjust its algorithm accordingly.

Customer Targeting and Segmentation

For the marketing companies to be able to reach their customers with a high level of personalization, they are required to target increasingly granular segments. The artificial intelligence technology can draw on the existing customer data and train Machine learning algorithms against "gold standard" training sets to identify common properties and significant variables. The data segments could be as simple as location, gender, and age, or as complex as the buyer's persona and past behavior. With AI, Dynamics Segmentation is feasible which accounts for the fact that customers' behaviors are ever-changing, and people can take on different personas in different situations.

Sales and Marketing Forecast

One of the most straightforward artificial intelligence applications in marketing is in the development of sales and marketing forecasting models. The high volume of quantifiable data such as clicks, purchases, email responses, and time spent on webpages serve as training resources for the machine learning algorithms. Some of the leading business intelligence and production companies in the market are Sisense, Rapidminer, and Birst. Marketing companies are continuously upgrading their marketing efforts, and with the help of AI and machine learning, they can predict the success of their marketing initiatives or email campaigns. Artificial intelligence technology can analyze past sales data, economic trends as well as industrywide comparisons to predict short and long-term sales performance, and forecast sales outcomes. The sales forecasts model aid in the estimation of product demand and to help companies manage their production to optimize sales.

Programmatic Advertisement Targeting

With the introduction of artificial intelligence technology, bidding on and targeting program based advertisement has become significantly more efficient. Programmatic advertising can be defined as "the automated process of buying and selling ad inventory to an exchange which connects advertisers to publishers." To allow real-time bidding for inventory across social media channels and mobile devices as well as television, artificial intelligence technology is used. This also goes back to predictive analysis and the ability to model data that could previously only be determined retroactively. Artificial intelligence is able to assist the best time of the day to serve a particular ad, the probability of an ad turning into sales, the receptiveness of the user, and the likelihood of engagement with the ad. Programmatic companies can gather and analyze visiting customers' data and behaviors to optimize real-time campaigns and to target the audience more precisely. Programmatic media buying includes the use of "demand-side platforms" (to facilitate the process of buying ad inventory on the open market) and "data management platforms" (to provide the marketing company an ability to reach their target audience). In order to empower the marketing rep to make informed decisions regarding their prospective customers, the data management platforms are designed to collect and analyze the big volume of website "cookie data." For example, search engine marketing (SEM) advertising is practiced by corporations like Facebook, Twitter, and Google. To efficiently manage a huge inventory of the website and application viewers, programmatic ads provide a significant edge over competitors. Google and Facebook serve as the gold standard for efficient and effective advertising and are geared to words providing a user-friendly platform that will allow nontechnical marketing companies to start, run and measure their initiatives and campaigns online.

Visual Search and Image Recognition

Leaps and bounds of the advancements in artificial intelligence-based image recognition and analysis technology have resulted in uncanny visual search functionalities. With the introduction of technology like Google Lens and platforms like Pinterest, people can now find results that are visually similar to one another using visual search functionality. The visual search works in the same way as traditional text-based searches that display results on a similar topic. Major retailers and marketing companies are increasingly using the visual search to offer an enhanced and more engaging customer experience. Visual search can be used to improve merchandising and provide product recommendations based on the style of the product instead of the consumer's past behavior or purchases.

Major investments have been made by Target and Asos in the visual search technology development for their e-commerce website. In 2017, Target announced a partnership with interest that allows integration of Pinterest's visual search application called "Pinterest lens" into Target's mobile application. As a result, shoppers can take a picture of products that they would like to purchase while they are out and about and find similar items on Target's e-commerce site. Similarly, the visual search application launched by Asos called "Asos' Style Match" allows shoppers to snap a photo or upload an image on the Asos website or application and search their product catalog for similar items. These tools attract shoppers to retailers for items that they might come across in a magazine or while out and about by helping them to shop for the ideal product even if they do not know what the product is.

Image recognition has tremendously helped marketing companies to gain an edge on social media by allowing them to find a variety of uses of their brand logos and products in keeping up with the visual trends. This phenomenon is also called "visual social listening" and allows companies to identify and understand where and how customers are interacting with their brand, logo, and product even when the company is not referred directly by its name.

Healthcare Industry

With the increasing availability of healthcare data, big data analysis has brought on a paradigm shift to healthcare. The primary focus of big data analytics in the healthcare industry is the analysis of relationships between patient outcomes and the treatment or prevention technique used. Big data analysis driven Artificial Intelligence programs have successfully been developed for patient diagnostics, treatment protocol generation, drug development, as well as patient monitoring and care. The powerful AI techniques can sift through a massive amount of clinical data and help unlock clinically relevant information to assist in decision making.

Some medical specialties with increasing big data analysis based AI research and applications are:

•   Radiology – The ability of AI to interpret imaging results supplements the clinician's ability to detect changes in an image that can easily be missed by the human eye. An AI algorithm recent created at Stanford University can detect specific sites in the lungs of the pneumonia patients.

•   Electronic Health Records – The need for digital health records to optimize the information spread and access requires fast and accurate logging of all health-related data in the systems. A human is prone to errors and may be affected by cognitive overload and burnout. This process has been successfully automated by AI. The use of Predictive models on the electronic health records data allowed the prediction of individualized treatment response with 70-72% accuracy at baseline.

•   Imaging – Ongoing AI research is helping doctors in evaluating the outcome of corrective jaw surgery as well as in assessing the cleft palate therapy to predict facial attractiveness.

Entertainment Industry

Big data analysis, in coordination with Artificial intelligence, is increasingly running in the background of entertainment sources from video games to movies and serving us a richer, engaging, and more realistic experience. Entertainment providers such as Netflix and Hulu are using big data analysis to provide users personalized recommendations derived from individual user's historical activity and behavior. Computer graphics and digital media content producers have been leveraging big data analysis based tools to enhance the pace and efficiency of their production processes. Movie companies are increasingly using machine learning algorithms in the development of film trailers and advertisements as well as pre-and post-production processes. For example, big data analysis and an artificial intelligence-powered tool called "RivetAI" allows producers to automate and excellently read the processes of movie script breakdown, storyboard as well as budgeting, scheduling, and generation of shot-list. Certain time-consuming tasks carried out during the post-production of the movies such as synchronization and assembly of the movie clips can be easily automated using artificial intelligence.

Marketing and Advertising

A machine learning algorithm developed as a result of big data analysis can be easily trained with texts, stills, and video segments as data sources. It can then extract objects and concepts from these sources and recommend efficient marketing and advertising solutions. For example, a tool called "Luban" was developed by Alibaba that can create banners at lightning speed in comparison to a human designer. In 2016, for the Chinese online shopping extravaganza called "Singles Day," Luban generated a hundred and 17 million banner designs at a speed of 8000 banner designs per second. The "20th Century Fox" collaborated with IBM to use their AI system "Watson" for the creation of the trailer of their horror movie "Morgan." To learn the appropriate "moments" or clips that should appear in a standard horror movie trailer, Watson was trained to classify and analyze input "moments" from audio-visual and other composition elements from over a hundred horror movies. This training resulted in the creation of a six-minute movie trailer by Watson in a mere 24 hours, which would have taken human professional weeks to produce. With the use of Machine learning, computer vision technology, natural language processing, and predictive analytics, the marketing process can be accelerated exponentially through an AI marketing platform. For example, the artificial intelligence-based marketing platform developed by Albert Intelligence Marketing can generate autonomous campaign management strategies, create custom solutions and perform audience targeting. The company reported a 183% improvement in customer transaction rate and over 600% higher conversation efficiency credited to the use of their AI-based platform.

In March 2016, the artificial intelligence-based creative director called "AI-CD ß" was launched by McCann Erickson Japan as the first robotic creative director ever developed. "AI-CD ß" was given training on select elements of various TV shows and the winners from the past 10 years of All Japan Radio and Television CM festival. With the use of data mining capabilities, "AI-CD ß" can extract ideas and themes fulfilling every client's individual campaign needs.

Security

There are several cities throughout the world that are working on predictive analysis so that they can predict the areas of the town where there is more likely to be a big surge of crimes. This is done with the help of some data from the past and even data on the geography of the area.

This is actually something that a few cities in America have been able to use, including Chicago. Although we can imagine that it is impossible to use this to catch every crime that is out there, the data that is available from using this is going to make it easier for police officers to be present in the right areas at the right times to help reduce the rates of crime in some of those areas. And in the future, you will find that when we use data analysis in this kind of manner in the big cities has helped to make these cities and these areas a lot safer, and the risks would not have to put their lives at risk as much as before.

Transportation

The world of transportation is able to work with data analysis, as well. A few years ago, when plans were being made at the London Olympics, there was a need during this event to handle more than 18 million journeys that were made by fans into the city of London. Moreover, it was something that we were able to sort out well.

How was this feat achieved for all of these people? The train operators and the TFL operators worked with data analytics to make sure that all those journeys went as smoothly as possible. These groups were able to go through and input data from the events that happened around that time and then used this as a way to forecast how many people would travel to it. This plan went so well that all of the spectators and the athletes could be moved to and from the right places in a timely manner the whole event.

Risk and Fraud Detection

This was one of the original uses of data analysis and was often used in the field of finance. There are many organizations that had a bad experience with debt, and they were ready to make some changes to this. Because they had a hold on the data that was collected each time that the customer came in for a loan, they were able to work with this process in order to not lose as much money in the process.

This allowed the banks and other financial institutions to dive and conquer some of the data from the profiles they could use from those customers. When the bank or financial institution is able to utilize their customers they are working with, the costs that had come up recently, and some of the other information that is important for these tools, they will make some better decisions about who to loan out money to, reducing their risks overall. This helps them to offer better rates to their customers.

In addition to helping these financial institutions make sure that they can hand out loans to customers who are more likely to pay them back, you will find that this can be used in order to help cut down on the risks of fraud as well. This can cost the bank billions of dollars a year and can be expensive to work with. When the bank can use all of the data that they have for helping discover transactions that are fraudulent and making it easier for their customers to keep money in their account, and make sure that the bank is not going to lose money in the process as well.

Logistics of Deliveries

There are no limitations when it comes to what we are able to do with our data analysis, and we will find that it works well when it comes to logistics and deliveries. There are several companies that focus on logistics, which will work with this data analysis, including UPS, FedEx, and DHL. They will use data in order to improve how efficient their operations are all about.

From applications of analytics of the data, it is possible for these companies who use it to find the best and most efficient routes to use when shipping items, the ones that will ensure the items will be delivered on time, and so much more. This helps the item to get things through in no time and keeps costs down to a minimum as well. Along with this, the information that the companies are able to gather through their GPS can give them more opportunities in the future to use data science and data analytics.

Customer Interactions

Many businesses are going to work with the applications of data analytics in order to have better interactions with their customers. Companies can do a lot about their customers, often with some customer surveys. For example, many insurance companies are going to use this by sending out customer surveys after they interact with their handler. The insurance company is then able to use which of their services are good, that the customers like, and which ones they would like to work on to see some improvements.

There are many demographics that a business is able to work with and it is possible that these are going to need many diverse methods of communication, including email, phone, websites, and in-person interactions. Taking some of the analysis that they can get with the demographics of their customers and the feedback that comes in, it will ensure that these insurance companies can offer the right products to these customers, and it depends one hundred percent on the proven insights and customer behavior as well.

Healthcare

The healthcare industry has been able to see many benefits from data analysis. There are many methods, but we are going to look at one of the main challenges that hospitals are going to face. Moreover, this is that they need to cope with cost pressures when they want to treat as many patients as possible while still getting high-quality care to the patients. This makes the doctors and other staff fall behind in some of their work on occasion, and it is hard to keep up with the demand.

You will find that the data we can use here has risen so much, and it allows the hospital to optimize and then track the treatment of their patient. It is also a good way to track the patient flow and how the different equipment in the hospital is being used. In fact, this is so powerful that it is estimated that using this data analytics could provide a 1 percent efficiency gain, and could result in more than $63 billion in worldwide healthcare services. Think of what that could mean to you and those around you.

Doctors are going to work with data analysis in order to provide them with a way to help their patients a bit more. They can use this to make some diagnoses and understand what is going on with their patients in a timely and more efficient manner. This can allow doctors to provide their customers with a better experience and better care while ensuring that they can keep up with everything they need to do.

Travel

Data analytics and some of their applications are a good way to help optimize the buying experience for a traveler. This can be true through a variety of options, including data analysis of mobile sources, websites, or social media. The reason for this is because the desires and the preferences of the customer can be obtained from all of these sources, which makes companies start to sell out their products thanks to the correlation of all the recent browsing on the site and any of the currency sells to help purchase conversions. They are able to utilize all of this to offer some customized packages and offers. The applications of data analytics can also help to deliver some personalized travel recommendations, and it often depends on the outcome that the company is able to get from their data on social media.

Travel can benefit other ways when it comes to working with the data analysis. When hotels are trying to fill up, they can work with data analysis to figure out which advertisements they would like to offer to their customers. Moreover, they may try to utilize this to help figure out which nights, and which customers, will fill up or show up. Pretty much all of the different parts of the travel world can benefit when it comes to working with data analysis.

Digital Advertising

Outside of just using it to help with some searching, there is another area where we are able to see data analytics happen regularly, and this is digital advertisements. From some of the banners that are found on several websites to the digital billboards that you may be used to seeing in some of the bigger and larger cities, but all of these will be controlled thanks to the algorithms of our data along the way.

This is a good reason why digital advertisements are more likely to get a higher CTR than the conventional methods that advertisers used to rely on a lot more. The targets are going to work more on the past behaviors of the users, and this can make for some good predictions in the future.

The importance that we see with the applications of data analytics is not something that we can overemphasize because it is going to be used in pretty much any and all of the areas of our life to ensure we have things go a bit easier than before. It is easier to see now, more than ever, how having data is such an important thing because it helps us to make some of the best decisions without any issues. However, if we don't have that data or we are not able to get through it because it is a mess and doo many points to look at, then our decisions are going to be based on something else. Data analysis ensures that our decisions are well thought out, that they make sense, and that they will work for our needs.

You may also find that when we inefficiently handle our data, it could lead to a number of problems. For example, it could lead to some of the departments that are found in a larger company so that we have a better idea of how we can use the data and the insights that we are able to find in the process, which could make it so that the data you have is not able to be used to its full potential. Moreover, if this gets too bad, then it is possible that the data will not serve any purpose at all.

However, you will find that as data is more accessible and available than ever before, and therefore more people, it is no longer just something that the data analysts and the data scientists are able to handle and no one else. Proper use of this data is important, but everyone can go out there and find the data they want. Moreover, this trend is likely to continue long into the future as well.

Chapter 3: Setting Up the Environment

Installation Instructions for Python

WINDOWS

1.   From the official Python website, click on the "Downloads" icon and select Windows.
2.   Click on the "Download Python 3.8.0" button to view all the downloadable files.
3.   On the subsequent screen, select the Python version you would like to download. We will be using the Python 3 version under "Stable Releases." So scroll down the page and click on the "Download Windows x86-64 executable installer" link, as shown in the picture below.
1.   A pop-up window titled "python-3.8.0-amd64.exe" will be shown.
2.   Click on the "Save File" button to start downloading the file.
3.   Once the download has completed, double click the saved file icon, and a "Python 3.8.0 (64-bit) Setup" pop window will be shown.
4.   Make sure that you select the "Install Launcher for all users (recommended)" and the "Add Python 3.8 to PATH" checkboxes. Note – If you already have an older version of Python installed on your system, the "Upgrade Now" button will appear instead of the "Install Now" button, and neither of the checkboxes will be shown.
5.   Click on "Install Now" and a "User Account Control" pop up window will be shown.
6.   A notification stating, "Do you want to allow this app to make changes to your device" will be shown, click on Yes.
7.   A new pop up window titled "Python 3.8.0 (64-bit) Setup" will be shown containing a setup progress bar.
8.   Once the installation has been completed, a "Set was successful" message will be shown. Click on Close.
9.   To verify the installation, navigate to the directory where you installed Python and double click on the python.exe file.

MACINTOSH

1.   From the official Python website, click on the "Downloads" icon and select Mac.

2.   Click on the "Download Python 3.8.0" button to view all the downloadable files.

3.   On the subsequent screen, select the Python version you would like to download. We will be using the Python 3 version under "Stable Releases." So scroll down the page and click on the "Download macOS 64-bit installer" link under Python 3.8.0, as shown in the picture below.

4.   A pop up window titled "python-3.8.0-macosx10.9.pkg" will be shown.

5.   Click "Save File" to start downloading the file.

6.   Once the download has completed, double click the saved file icon, and an "Install Python" pop window will be shown.

7.   Click "Continue" to proceed, and the terms and conditions pop up window will appear.

8.   Click Agree and then click "Install."

9.   A notification requesting administrator permission and password will be shown. Enter your system password to start the installation.

10.   Once the installation has finished, an "Installation was successful" message will appear. Click on the Close button, and you are all set.

11.   To verify the installation, navigate to the directory where you installed Python and double click on the python launcher icon that will take you to the Python Terminal.

Getting Started

With the Python terminal installed on your computer, you can now start writing and executing the Python code. All Python codes are written in a text editor as (.py) files and executed on the Python interpreter command line as shown in the code below, where "nineplanets.py" is the name of the Python file:
"C: \Users\Your Name\python nineplanets.py"
You will be able to test a small code without writing it in a file and simply executing it as a command line itself by typing the code below on the Mac, Windows or Linux command line, as shown below:
"C: \Users\Your Name\python"
In case the command above does not work, use the code below instead:
"C: \Users\Your Name\py"

Indentation – The importance of indentation, which is the number of spaces preceding the code, is fundamental to the Python coding structure. In most programming languages, indentation is added to enhance the readability of the code. However, in Python, the indentation is used to indicate the execution of a subset of the code, as shown in the code below

If 7 > 2:

print ('Seven is greater than two')

Indentation precedes the second line of code with the print command. If the indentation is skipped and the code was written as below, an error will be triggered:

If 7 > 2:

print ('Seven is greater than two')

The number of spaces can be modified but is required to have at least one space. For example, you can execute the code below with higher indentation, but for a specific set of code same number of spaces must be used, or you will receive an error.

If 7 > 2:

print ('Seven is greater than two')

Adding Comments – In Python, comments can be added to the code by starting the code comment lines with a "#," as shown in the example below:

#Any relevant comments will be added here

print ('Nine planets')

Comments serve as a description of the code and will not be executed by the Python terminal. Make sure to remember that any comments at the end of the code line will lead to the entire code line being skipped by the Python terminal, as shown in the code below. Comments can be very useful in case you need to stop the execution when you are testing the code.

print ('Nine Planets')#Comments added here

Multiple lines of comments can be added by starting each code line with "#," as shown below:

#Comments added here

#Supplementing the comments here

#Further adding the comments here

print ('Nine Planets')

Python Variables

In Python, variables are primarily utilized to save data values without executing a command for it. A variable can be created by simply assigning the desired value to it, as shown in the example below:

```
A = 999
B = 'Patricia'
print (A)
print (B)
```

A variable could be declared without a specific data type. The data type of a variable can also be modified after it's an initial declaration, as shown in the example below:

```
A = 999 # A has data type set as int
A = 'Patricia' # A now has data type str
```

```
print (A)
```

Some of the rules applied to the Python variable names are as follows:

1.   Variable names could be as short as single alphabets or more descriptive words like height, weight, and more.
2.   Variable names could only be started with an underscore character or a letter.
3.   Variable names must not start with numbers.
4.   Variable names can contain underscores or alphanumeric characters. No other special characters are allowed.
5.   Variable names are case sensitive. For example, 'weight,' 'Weight,' and 'WEIGHT' will be accounted for as 3 separate variables.

Assigning Value to Variables

In Python, multiple variables can be assigned DISTINCT values in a single code line, as shown in the example below:

A, B, C = 'violet,' maroon, 'teal'

print (A)

print (B)

print (C)

OR multiple variables can be assigned SAME value in a single code line, as shown in the example below:

A, B, C = 'violet'

print (A)

print (B)

print (C)

To view the data type of any object, you can use the "type ()" function as shown in the example below:
A = 'Violet'
print (type (A))

Assigning the Data Type to Variables

A new variable can be created by simply declaring a value for it. This set data value will in turn assign the data type to the variable.
To assign a specific data type to a variable, the constructor functions listed below are used:

Constructor Functions	Data Type
A = str ('Nine Planets)'	str
A = int (55)	Int (Must be a whole number, positive or negative with no decimals, no length restrictions)
A = float (14e6)	Float

	(Floating point number must be positive or negative number with one or more decimals; maybe scientific number an 'e' to specify an exponential power of 10)
A = complex (92j)	Complex (Must be written with a 'j' as an imaginary character)
A = list (('teal', maroon, 'jade'))	list
A = range (3, 110)	range
A = tuple (('teal', maroon, 'jade'))	tuple
A = set (('teal', maroon, 'jade'))	set
A = frozenset (('teal', 'jade', maroon))	frozenset
A = dict ('color' : maroon, 'year' : 1988)	dict
A = bool (False)	bool
A = bytes (542)	bytes
A = bytearray (9)	bytearray
A = memoryview (bytes (525))	memoryview

EXERCISE – To solidify your understanding of data types; look at the first column of the table below and write the data type for that variable. Once you have all your answers, look at the second column, and verify your answers.

Variable	Data Type
A = 'Nine Planets'	str
A = 45	int
A = 56e2	float
A = 34j	complex
A = ['teal', maroon, 'jade']	list
A = range (12, 103)	range
A = ('teal', maroon,	tuple

'jade')	
A = {'teal', maroon, 'jade'}	set
A = frozenset ({ 'teal', 'jade', maroon})	frozenset
A = ['color' : maroon, 'year' : 1939}	dict
A = False	bool
A = b 'Morning'	bytes
A = bytearray (5)	bytearray
A = memoryview (bytes (45))	memoryview

Output Variables

In order to retrieve variables as output, the "print" statements are used in Python. You can use the "+" character to combine text with a variable for final output, as shown in the example below:

'A = maroon

print ('Flowers are' + A)'

OUTPUT – 'Flowers are maroon'

A variable can also be combined with another variable using the "+" character as shown in the example below:

'A = 'Flowers are'

B = maroon

AB = A + B

print (AB)'

OUTPUT – 'Flowers are maroon'

However, when the "+" character is used with numeric values, it retains its function as a mathematical operator, as shown in the example below:

```
'A = 22
B = 33
print (A + B)'
OUTPUT = 55
```

You will not be able to combine a string of characters with numbers and will trigger an error instead, as shown in the example below:

```
A = yellow
B = 30
print (A + B)
OUTPUT – N/A – ERROR
```

Chapter 4: Python Data Structures Essentials

Python Data Structures – Lists, Tuples, Sets, Dictionaries

A data structure is a way to organize and store data where we can access and modify it efficiently.
Let's begin with our first Python Data Structures and lists.

Python List

In Python, lists are collections of data types that can be changed, organized, and include duplicate values. Lists are written within square brackets, as shown in the syntax below.

X = ["string001", "string002", "string003"]

print (X)

The same concept of position applies to Lists as the string data type, which dictates that the first string is considered to be at position 0. Subsequently, the strings that will follow are given positions 1, 2, and so on. You can selectively display desired string from a List by referencing the position of that string inside square bracket in the print command, as shown below.

X = ["string001", "string002", "string003"]

print (X [2])

OUTPUT – [string003]

Similarly, the concept of negative indexing is also applied to Python List. Let's look at the example below:

```
X = ["string001", "string002", "string003"]
print (X [-2])
```

OUTPUT – [string002]

You will also be able to specify a range of indexes by indicating the start and end of a range. The result in values of such command on a Python List would be a new List containing only the indicated items. Here is an example for your reference.

```
X = ["string001", "string002", "string003", "string004",
"string005", "string006"]
print (X [2 : 4])
```

OUTPUT – ["string003", "string004"]

* Remember the first item is at position 0, and the final position of the range (4) is not included.

Now, if you do not indicate the start of this range, it will default to position 0 as shown in the example below:

```
X = ["string001", "string002", "string003", "string004",
"string005", "string006"]
print (X [ : 3])
```

OUTPUT – ["string001", "string002", "string003"]

Similarly, if you do not indicate the end of this range it will display all the items of the List from the indicated start range to the end of the List, as shown in the example below:

```
X = ["string001", "string002", "string003", "string004",
"string005", "string006"]
print (X [3 : ])
```

OUTPUT – ["string004", "string005", "string006"]

You can also specify a range of negative indexes to Python Lists, as shown in the example below:

```
X = ["string001", "string002", "string003", "string004", "string005", "string006"]

print (X [-3 : -1])
```

OUTPUT – ["string004", "string005"]

* Remember the last item is at position -1, and the final position of this range (-1) is not included in the Output.

How to Declare Python List?

Just input:

```
>>> languages=['C++','Python','Scratch']
```
You can put any kind of value in a list.
```
>>> list1=[1,[2,3],(4,5),False,'No']
```

How to Access Python List?

Here's what you need to do to access an entire list, just type its name in the shell.

```
>>> list1
[1, [2, 3], (4, 5), False, 'No']
```

Slicing

If you would like to view a range of characters, you can do so by specifying the start and the end index of the desired positions and separating the indexes by a colon. For example, to view the characters of a string from position 2 to position 5, your code will be "print (variable [2:5])."

You can even view the characters starting from the end of the string by using "negative indexes" and start slicing the string from the end of the string. For example, to view characters of a string from position 3 to position 1, your code will be
"print (variable [-3: -2])."

In order to view the length of the string, you can use the "len ()" function. For example, to view the length of a string, your code will be "print (len (variable))."

How to Delete a Python List?

To remove a list element, you can use either the del statement if you know exactly which element(s) you are deleting or the remove() method if you do not know

Delete all the list

```
>>> a=[2,3,4]
>>> del a
```

Delete One element of the list

```
>>> a=[2,3,4]
>>> del a[0]
```

Python Tuple

In Python, Tuples are collections of data types that cannot be changed but can be arranged in specific order. Tuples allow for duplicate items and are written within round brackets, as shown in the syntax below.

```
Tuple = ("string001", "string002", "string003")
print (Tuple)
```

Similar to the Python List, you can selectively display the desired string from a Tuple by referencing the position of that string inside the square bracket in the print command as shown below.

```
Tuple = ("string001", "string002", "string003")
print (Tuple [1])
```

OUTPUT – ("string002")

Accessing, Reassigning, and Deleting Items

Unlike Python lists, you cannot directly change the data value of Python Tuples after they have been created. However, conversion of a Tuple into a List and then modifying the data value of that List will allow you to subsequently create a Tuple from that updated List. Let's look at the example below:

```
Tuple1 = ("string001", "string002", "string003", "string004", "string005", "string006")
List1 = list (Tuple1)
List1 [2] = "update this list to create new tuple"
Tuple1 = tuple (List1)
print (Tuple1)
```

OUTPUT – ("string001", "string002", "update this list to create new tuple", "string004", "string005", "string006")
You can also determine the length of a Python Tuple using the "len()" function, as shown in the example below:

```
Tuple = ("string001", "string002", "string003", "string004", "string005", "string006")
print (len (Tuple))
```

You cannot selectively delete items from a Tuple, but you can use the "del" keyword to delete the Tuple in its entirety, as shown in the example below:
Tuple = ("string001", "string002", "string003", "string004")
del Tuple

print (Tuple)

OUTPUT – name 'Tuple' is not defined

You can join multiple Tuples with the use of the "+" logical operator.

Tuple1 = ("string001", "string002", "string003", "string004")
Tuple2 = (100, 200, 300)

Tuple3 = Tuple1 + Tuple2
print (Tuple3)

OUTPUT – ("string001", "string002", "string003", "string004", 100, 200, 300)

You can also use the "tuple ( )" constructor to create a Tuple, as shown in the example below:

Tuple1 = tuple (("string001", "string002", "string003", "string004"))
print (Tuple1)

Python Set

In Python, sets are collections of data types that cannot be organized and indexed. Sets do not allow for duplicate items and must be written within curly brackets, as shown in the syntax below:

```
set = {"string1", "string2", "string3"}
print (set)
```

Unlike the Python List and Tuple, you cannot selectively display desired items from a Set by referencing the position of that item because the Python Set are not arranged in any order. Therefore, items do not have any indexing. However, the "for" loop can be used on Sets (more on this topic later in this chapter).

Unlike Python Lists, you cannot directly change the data values of Python Sets after they have been created. However, you can use the "add ()" method to add a single item to Set and use the "update ( )" method to one or more items to an already existing Set. Let's look at the example below:

```
set = {"string1", "string2", "string3"}
set. add ("newstring")
print (set)
```

OUTPUT – {"string1", "string2", "string3", "newstring"}

```
set = {"string1", "string2", "string3"}
set. update (["newstring1", "newstring2", "newstring3".)
print (set)
```

OUTPUT – {"string1", "string2", "string3", "newstring1", "newstring2", "newstring3"}

Python Dictionaries

Python dictionary is used to store data by using a unique key for each entry. Key is somehow serving the same functionality as indices in lists and tuples when we want to refer to a certain element. The values where each key refers to are not necessarily unique. Diction-

aries are mutable, so we can add, delete or modify the elements in them.
The entries of a dictionary are contained between curly braces {}. Each entry is in the form key: value.

```
>>> mydict={1:2,2:4,3:6}

>>> mydict
```

Chapter 5: IPython/Jupyter Notebook - Interactive Computational Environment

Getting started with Jupyter Notebook (IPython)

The Jupyter Notebook is an open-source web application that permits you to produce and share files that contain live code, formulas, visualizations, and narrative text. Utilizes consist of information cleansing and change, mathematical simulation, analytical modeling, information visualization, artificial intelligence, and far more.

Jupyter has assistance for over 40 various shows languages and Python is among them. Python is a requirement (Python 3.3 or higher, or Python 2.7) for setting up the Jupyter Notebook itself.

Setting up Jupyter utilizing Anaconda

Set Up Python and Jupyter utilizing the Anaconda Distribution, which includes Python, the Jupyter Notebook, and other typically utilized bundles for clinical computing and information science. You can download Anaconda's newest Python3 variation.

Now, set up the downloaded variation of Anaconda. Setting up Jupyter Notebook utilizing PIP:

```
python3 -m pip install --upgrade pip
```

```
python3 -m pip install jupyter
```

Command to run the Jupyter notebook:

```
jupyter notebook
```

This will print some details about the note pad server in your terminal, consisting of the URL of the web application, by default: http://localhost:8888

and after that open your default Web Internet browser to this URL.

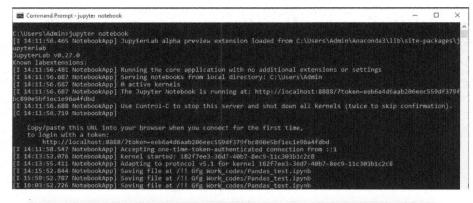

When the Notepad opens in your Internet browser, you will see the Notebook Dashboard, which will reveal a list of the notepads, files, and subdirectories in the directory site where the Notepad server was started. The majority of the time, you will want to begin a Notepad server in the greatest level directory site consisting of notepads. Typically this will be your house directory site.

Create a new Notebook

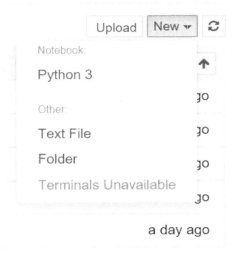

Now on the control panel, you can see a brand-new button on the top right corner. Click it to open a drop-down list and after that, if you'll click Python3, it will open a brand-new notebook.

Useful Commands

Command to open a notebook in the currently running notebook server. jupyter notebook

notebook_name.ipynb

By default, the note pad server begins on port 8888. If port 8888 is not available or in usage, the Notepad server browses the next readily available port.

jupyter note pad-- port 9999

Command to begin the Notepad server without opening a Web Internet browser:

```
jupyter notebook --no-browser
```

The notebook server provides help messages for other command-line arguments using the help flag:

```
jupyter notebook --help
```

Running your First code in Jupyter

Action #1: After effectively setting up Jupyter compose 'jupyter note pad' in the terminal/command timely. This will open a brand-new Notepad server on your Web Internet browser.

Action #2: On the leading left corner, click the brand-new button and choose python3. This will open a brand-new Notepad tab in your Web browser where you can begin to compose your very first code.

Action #3: Press Click or get in on the very first cell in your Notepad to enter into the edit mode.

Action #4: Now you are free to compose any code.

Action #5: You can run your code by pushing Shift + Enter or the run button offered at the top.

```
In [6]:   # Define a function for addition
          def add(a , b):
              return a + b

          val = add(12, 13)

          val

Out[6]:   25
```

Jupyter Notebook Tips and Tricks

Python is a fantastic language for doing information analysis, mainly because of the great environment of data-centric python bundles. What makes information analysis in Python more efficient and effective is Jupyter Notepad or what was previously called the IPython Notepad.

In this area, we are going to go over some great functions of the Jupyter Notepad which increases the efficiency and effectiveness of the information expert. The Jupyter Notepad extends the console-based technique to interactive computing in qualitatively brand-new instructions, offering a web-based application appropriate for recording the entire calculation procedure: establishing, recording, and carrying out code, along with interacting with the outcomes. In a nutshell, it is a total plan.

Let's see some functions of the Jupyter Notepad which becomes extremely convenient while doing information analysis.

%%timeit and %%time

It's not an unusual thing for an information researcher that while doing information analysis, they have more than one service for the provided issue. They wish to pick the very best technique which finishes the job in the minimum quantity of time. Jupyter Notepad supplies an extremely effective method to inspect the running time of a specific block of code.

We can utilize the %% time command to inspect the running time of a specific cell. Let's see the time takes to carry out the code pointed out listed below.

```
# For capturing the execution time

%%time
# Find the squares of a number in the # range from 0 to 14

for x in range(15):
square = x**2 print(square)
```

```
0
1
4
9
16
25
36
49
64
81
100
121
144
169
196
Wall time: 999 µs
```

Commenting/Uncommenting a block of code

While dealing with codes, we typically include brand-new lines of code and comment out the old pieces of code for enhancing the efficiency or to debug it. Jupyter Notepad supplies an extremely effective method to accomplish the same – to comment out a block of code.

```
1  df = pd.DataFrame({'Date':['10/2/2011','11/2/2011','12/2/2011','13/2/2011'],
2                     'Event':['Music','Poetry','Theatre','Comedy'],
3                     'Cost':[10000,5000,15000,2000]})
4
5  df.index = ['A', 'B', 'A', 'D']
6
7  print(df)
8
9  df2 = pd.DataFrame({'Cost1':[10025,5700,2415,1800],
10                     'Cost2':[1000,500,150,20],
11                     'Cost3':[10000,5000,15000,2000]})
12
13 print(df.columns)
```

We require to choose all those lines which we want to comment out, as shown in the following picture:

```
|:  1  df = pd.DataFrame({'Date':['10/2/2011','11/2/2011','12/2/2011','13/2/2011'],
    2                    'Event':['Music','Poetry','Theatre','Comedy'],
    3                    'Cost':[10000,5000,15000,2000]})
    4
    5  df.index = ['A', 'B', 'A', 'D']
    6
    7  # print(df)
    8
    9  # df2 = pd.DataFrame({'Cost1':[10025,5700,2415,1800],
   10  #                    'Cost2':[1000,500,150,20],
   11  #                    'Cost3':[10000,5000,15000,2000]})
   12
   13  print(df.columns)
```

Next, on a Windows computer system, we need to push the ctrl +/ crucial mix to comment out the highlighted part of the code. This does conserve a great deal of time for the information expert.

```
:  1  df = pd.DataFrame({'Date':['10/2/2011','11/2/2011','12/2/2011','13/2/2011'],
   2                    'Event':['Music','Poetry','Theatre','Comedy'],
   3                    'Cost':[10000,5000,15000,2000]})
   4
   5  df.index = ['A', 'B', 'A', 'D']
   6
   7  print(df)
   8
   9  df2 = pd.DataFrame({'Cost1':[10025,5700,2415,1800],
  10                     'Cost2':[1000,500,150,20],
  11                     'Cost3':[10000,5000,15000,2000]})
  12
  13  print(df.columns)
```

Next, on a Windows computer system, we need to push the ctrl +/ crucial mix to comment out the highlighted part of the code.

Chapter 6: NumPy - Scientific Computing and Numerical Data Processing

What is A Python NumPy Array?

NumPy is one of the fundamental packages that are out there when you want to use Python to do scientific computing. It is going to be a Python library that you can choose to use because it will provide you with a lot of different things that make scientific computing a little bit easier to work with. First, it is going to provide you with an array of objects that are seen as multidimensional. Then it is able to provide us with various objects that are derived, which could include matrices and masked arrays. And then there is an assortment of routines that are used for making the operation on the array that much faster.

Another thing to consider is that the arrays that come with NumPy are going to be able to help out with some more of the advanced mathematical options that you are able to do, including operations that are meant to work on large numbers of data at the same time. While it is possible to do this with the Python sequence if you choose, it is going to take a lot less code and will be more efficient to rely on the arrays from NumPy instead.

How to Install NumPy

To get NumPy, you should download the Anaconda Python distribution. This is easy and will allow you to get started quickly! If you haven't downloaded it already, go get it. Follow the instructions to install, and you're ready to start! Do you wonder why this might actually be easier?

The good thing about getting this Python distribution is the fact that you don't need to worry too much about separately installing NumPy or any of the major packages that you'll be using for your data analyses, such as pandas, scikit-learn, etc. Because, especially if you're very new to Python, programming or terminals, it can really come as a relief that Anaconda already includes 100 of the most popular Python, R and Scala packages for data science. But also for more seasoned data scientists, Anaconda is the way to go if you want to get started quickly on tackling data science problems.

What's more, Anaconda also includes several open source development environments such as Jupyter and Spyder. In short, consider downloading Anaconda to get started on working with numpy and other packages that are relevant to data science!

How to Make NumPy Arrays

From here, we need to take some time to learn how to create these arrays. We will assume that you already have the NumPy library on your computer and ready to go. There are then two main ways that we are able to create some of these arrays including:

You can go through and make one of these arrays with the nested list or the Python list.

We can also work with some of the methods that are built-in with NumPy to make these arrays.

We are going to start out by looking at the steps that are necessary in order to create an array from the nested list and the Python list. To do this, we just need to pass the list from Python with the method of np.array() as your argument, and then you are done. When you do this, you will get either a vector or a 1D array, which can help you to get a lot of the necessary work done.

There are also times when we want to take this a bit further. We would want to get out of the 1D array that we just created, and we want to turn it into a 2D array or a matrix. To do this, we simply need to pass the Python list of lists to the method of np.array(), and then it is done for us.

How NumPy Broadcasting Works

Broadcasting refers to the ability of NumPy to perform arithmetic operations on different shaped arrays. If two arrays have the same shape, then arithmetic operations are easily performed since they are between corresponding elements. This becomes impossible in the case of differently sized arrays but NumPy has a solution for this. In NumPy the smaller array is "broadcasted" across the larger array so that they have the same shapes. It will be clear by this example:

Suppose I have an array 'A' of 4X5 shape and an array 'B' of size 1X5 then the smaller array B is stretched such that it also has a size of

4X5 same as 'A'. The new elements in B are simply a copy of the original elements.

1	71	8	5	0
7	4	35	8	11
62	0	4	14	2
2	15	86	13	10

Array A

1	3	2	1	5
1	3	2	1	5
1	3	2	1	5
1	3	2	1	5

Array B

2	74	10	6	5
8	7	37	9	16
63	3	6	15	7
3	18	88	14	15

Array A+b

Note: NumPy doesn't actually make copies of elements instead it uses the original scalar value repeatedly to make it as memory and computationally efficient as possible.

How to Slice, And Index Arrays

Sometimes you need only a subset of your data or need the value of some individual elements. Then you can do this by indexing and slicing. This is the way to address individual elements. One important thing to remember is that indexing always starts with '0'. For a single-dimensional matrix indexing is pretty easy.
The syntax for indexing in 1-dimensional matrix is given as arr[n], this gives you n-th number starting with zero indexes.
For a 2-dimensional matrix, you can pass a comma-separated list of indices to select individual elements. It is given as arr[m,n] or arr[m][n], this gives you the element on m-th row and nth column starting with zero indexes.

	0	1	2
0	(0,0)	(0,1)	(0,2)
1	(1,0)	(1,1)	(1,2)
2	(2,0)	(2,1)	(2,2)

Now when we talk about indexing in the multidimensional matrix, it becomes a little more difficult. Multidimensional matrix is basically a list of list with multiple layers.

Here I am going to talk about a 3-dimensional matrix. Indexing in a 3d array works like this: syntax: arr[L,M,N] or arr [L][M][N] where L is the first index, M is the row no. and N is the column no.

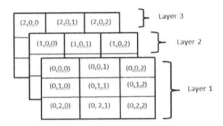

If you omit later indices in multidimensional arrays then all data along higher dimension will be returned for example in a 3d array if we search for arr[3] it return all the elements of layer 3.
All this is shown in example below:

>>> Array1= np.array([1, 2, 3, 4, 5, 6])

Array1[5]

6

>>> Array1=Array1.reshape(2,3)

>>> Array1[0,1]

2

>>>array3d =np.array([[[1, 2, 5],[6,0,2]] , [[5,9,8],[0,3,2]]])

>>> array1[1,0,1]

9

>>> array1[1]

array([[5, 9, 8]

[0, 3, 2]])

Slicing on the other hand is basically an operation that is used mainly to select a subset of an existing array. Continues values can be selected separated by colon (:). In all these cases the returned array are views. Views are basically virtual table that shows most recent data. However, it doesn't store the data. This means that the data is not copied, and any modifications will be reflected in the source array. So it's better to create a copy of that array if you feel like you will be needing the original data later.

Slicing in 1-D array: arr[x : y] , It returns all the elements indexed from x to (y-1) .

In multidimensional arrays, you can address it using rows and columns to include. It will be more clear to understand by visualizing it in this way:

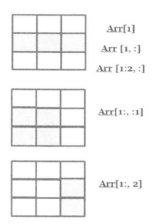

Arr[1]

Arr [1, :]

Arr [1:2, :]

Arr[1:, :1]

Arr[1:, 2]

Chapter 7: Matplotlib – Data Visualization

What is the Matplotlib library

Matplotlib is a library with which it is possible to draw 2d and 3d graphics. To use it, the first thing we need to do is import it, and in doing so we will take advantage of pyplot which offers a programming interface similar to MATLAB.

Import Matplotlib

```
> import matplotlib.pyplot as plt
```

we also import a very powerful numpy math library

```
> import numpy as np
```

Simple Graph

we choose a function to graph

```
> def straight_line (x):
>       return 4 * x + 1
```

we indicate that the xs can vary from 0 to 100

```
> x = np.arange (100)
```

we build the line

> straight y = (x)

let's draw the graph

> plt.plot (x, y)

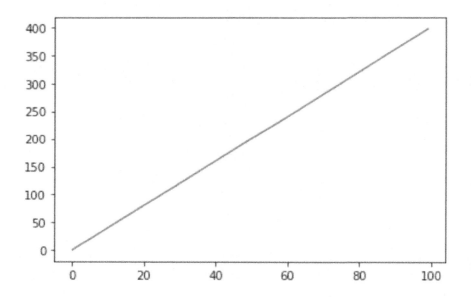

for completeness I report the example in jupyter.

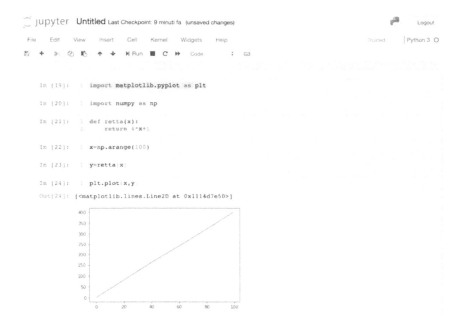

Complex Graphs

let's try to make the same example but with a slightly more complex graph

```
import matplotlib.pyplot as plt

import numpy as np

def f (x):

    return 20 + 105 * np.cos (x) * np.sin (x + 30) + 144 *
np.exp (0.5 / x + 1)

x = np.arange (-100,100)

y = f (x)
```

<u>plt.plot (x, y)</u>

we try to deepen the possibilities offered by pyplot.

To change the color of the graphs we can use the instruction:

<u>plt.plot (x, y, color = "<color_RGB>")</u>

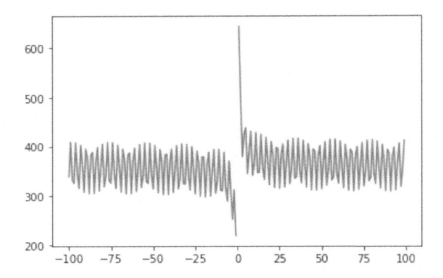

<u>plt.plot (x, y, color = '3412F3')</u>

to change the range of the axes I can use the instructions:

<u>plt.xlim (left_x, right_x) or plt.ylim (bottom_y, top_y)</u>

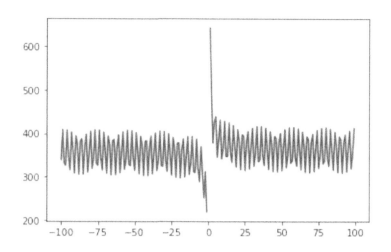

plt.xlim (-10.76)

plt.ylim (-100.600)

plt.plot (x, y)

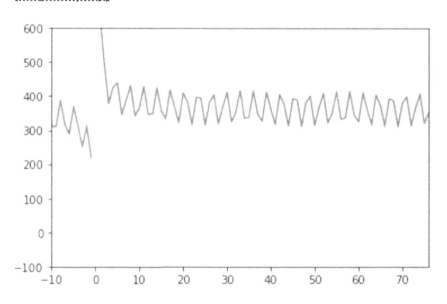

if we wanted, for example, to draw a graph with a dotted or dotted line we could use the parameter linestyle = "dashed" or linestyle = "dotted" for example:

plt.plot (x, y, linestyle = "dashed")

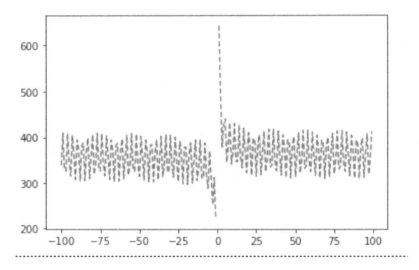

If we wanted to use markers we could use the marker = 'or' parameter

plt.plot (x, y, marker = 'o')

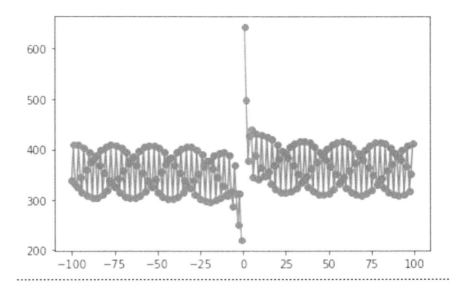

to add information to the axes or a legend we can use the following instructions:

plt.plot (x, y, label = "y = f (x)")

plt.title ("chart example")

plt.xlabel ("x course")

plt.ylabel ("trend of f (x)")

plt.legend ()

To draw more charts we can follow the following approach:

- we define the functions to be graphed,
- we define the x range
- We "plot" the graphs by repeating the plot command for the number of functions to be represented.

```python
import matplotlib.pyplot as plt
import numpy as np

def f (x):
    return 20 + 105 * np.cos (x) * np.sin (x + 30) + 144 * np.exp (0.5 / x + 1)

def straight line (x):
    return 4 * x + 1

x = np.arange (-100.100)
plt.plot (x, f (x))
plt.plot (x, straight line (x))
```

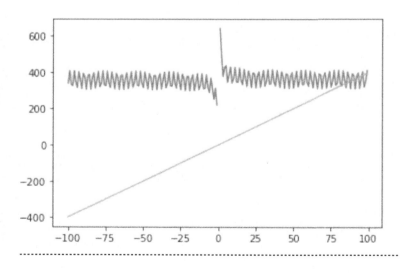

Obviously it is possible to use the various parameters in a single instruction, for example to represent a function of purple color, with circular markers and style of the dotted line we can write:

```
plt.plot (x, y, color = '# 8712F3', marker = 'o', linestyle = 'dashed', linewidth = 2, markersize = 12)
```

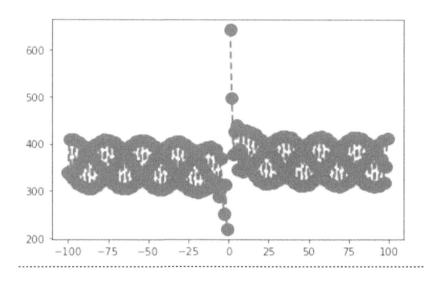

Scatter charts

first we import the necessary libraries:

```
import pandas as pd
import numpy as np
import matplotlib.pyplot as plt
```

To generate the points in the scatter space we use numpy.

#Generate random points

```
randomPoints = np.random.randn (100,2);
np.random.randn returns the samples in the scatter
space.
```

If integer positive arguments are supplied, randn generates an array (d0, d1, ..., dn), filled with random floats sampled according to a "normal" (Gaussian) distribution of mean 0 and variance 1. If no argument is given it will be returned a single random float sample of the standard distribution.

The previous instruction will generate 100 points randomly. The result will be something like this:

```
array ([[- 0.28393081, 1.69444015],
    [1.18710769, 0.75168393],
    [-1.08797599, 0.16593107],
    [0.34718794, 1.50584947],
    [0.59907203, 0.34574815],
    [-1.17150305, 1.32782198],
    [-0.66961386, 0.13941936],
    [0.96038988, -0.04362326],
    [-0.42828503, -1.35202516],
    [-0.57520301, -0.5182515]]])
```

At this point, we can create the dataframe that we will use as a parameter of pyplot matplotlib

```
df = pd.DataFrame (randomPoints, columns = ['x', 'y'])
```

and print the scatter chart

```
plt.plot (df ['x'], df ['y'], '.')
```

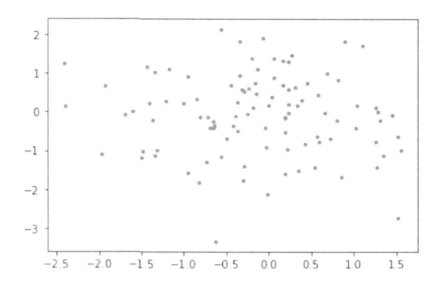

Complete example

```
In [17]:    1  import pandas as pd
            2  import numpy as np
            3  import matplotlib.pyplot as plt
            4  #Generiamo dei punti casuali
            5  randomPoints=np.random.randn(100,2);

In [18]:    1  df = pd.DataFrame(randomPoints,columns=['x', 'y'])

In [19]:    1  df.head()
```

Out[19]:

	x	y
0	-0.951031	0.925133
1	1.282889	0.020743
2	-0.045163	-0.401519
3	-0.659605	-0.225288
4	-1.502277	-1.175636

```
In [23]:    1  plt.plot(df['x'],df['y'],'.')
```

Out[23]: [<matplotlib.lines.Line2D at 0x118463050>]

If we wanted to represent more obvious points we could use the "o" instead of "."

```
plt.plot (df [ 'x'], df [ 'y'], 'o')
```

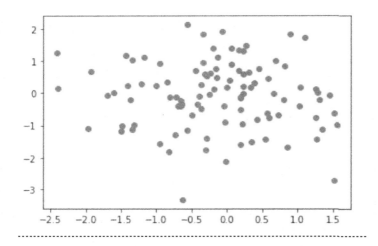

other options may allow us to create more significant graphs, defining the size of the points and the color, but to do this we will have to use the scatter function instead of the plot.

For our example we will use random data also for the colors and for the size of the points;

```
randomcolor = np.random.randn (100)
r_random_point= 1000*np.random.randn(100)
plt.scatter (df [ 'x'], df [ 'y'], s = r_random_point, alpha = 0.6, c = randomcolor)
```

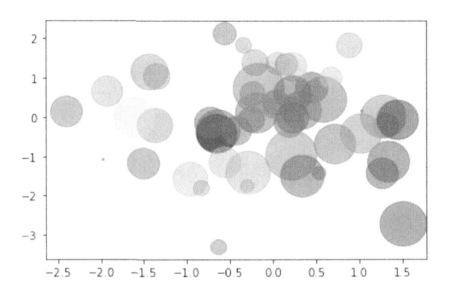

Chapter 8: Other Data Visualizatino Libraries

Seaborn

Seaborn outfits the intensity of matplotlib to make first-rate graphs in a couple of lines of code. The key difference is Seaborn's default patterns and shading palettes, which are meant to be all the greater tastefully pleasant and present day. Since Seaborn is primarily based over matplotlib, you will have to know matplotlib to trade Seaborn's defaults.

ggplot

ggplot relies upon on ggplot2, a R plotting framework, and ideas from The Grammar of Graphics. ggplot works uniquely in distinction to matplotlib: it gives you a hazard to layer components to make a whole plot. For example, you can begin with tomahawks, at that point include focuses, at that point a line, a trendline, and so forth. In spite of the fact that The Grammar of Graphics has been adulated as an "instinctive" approach for plotting, prepared matplotlib consumers may want time to acclimate to this new outlook.

As indicated by way of the maker, ggplot isn't supposed for making profoundly tweaked illustrations. It penances unpredictability for an extra straightforward method for plotting.

ggplot is firmly coordinated with pandas, so it's ideal to shop your statistics in a DataFrame when making use of ggplot.

Bokeh

Like ggplot, Bokeh relies upon on The Grammar of Graphics, alternatively no longer at all like ggplot, it is neighborhood to Python, no longer ported over from R. Its high-quality lies in the ability to make intuitive, web-prepared plots, which can be efficaciously yield as JSON objects, HTML records, or smart net applications. Bokeh likewise helps spilling and continuous information.

Bokeh furnishes three interfaces with transferring stages of control to oblige specific purchaser types. The most extended degree is for making outlines rapidly. It incorporates strategies for making regular graphs, for example, bar plots, field plots, and histograms. The middle level has a similar particularity as matplotlib and enables you to control the essential structure squares of each graph (the dabs in a disperse plot, for instance). The most decreased stage is designed for engineers and programming engineers. It has no pre-set defaults and expects you to characterize every element of the outline.

pygal

Like Bokeh and Plotly, pygal presents intuitive plots that can be inserted in the web browser. Its high differentiator is the capacity to yield outlines as SVGs. For something size of time that you are working with littler datasets, SVGs will do you best and dandy. However, in case you are making outlines with a large wide variety of information focuses, they'll journey issue rendering and come to be languid.
Since every outline kind is bundled into a method and the inherent styles are beautiful, it's whatever but tough to make a quality looking diagram in a couple of lines of code.

Plotly

You might also recognize Plotly as an on-line stage for statistics perception, but did you likewise realize you can get to its abilities from a Python scratchpad? Like Bokeh, Plotly's strong point is making wise plots, but it gives a few outlines you may not discover in many libraries, comparable to shape plots, dendograms, and 3D graphs.

geoplotlib

geoplotlib is a tool package for making maps and plotting geological information. You can make use of it to make an assortment of information types, as choropleths, heatmaps, and speck thickness maps. You should have Pyglet (an article located programming interface) added to utilize geoplotlib. Regardless, on the grounds that most Python data appreciation libraries don't provide maps, it's fantastic to have a library dedicated exclusively to them.

Glimmer

Glimmer is motivated by using R's Shiny bundle. It permits you to seriously change examinations into intuitive net applications utilizing simply Python contents, so you do not want to recognize some other dialects like HTML, CSS, or JavaScript. Glimmer works with any Python records representation library. When you've got made a plot, you can fabricate fields over it so clients can channel and kind information.

missingno

Managing missing information is a torment. missingno allows you to unexpectedly measure the fulfillment of a dataset with a visible outline, rather than on foot via a table. You can channel and kind data based on success or spot relationships with a heatmap or a dendrogram.

Calfskin

Calfskin's maker, Christopher Groskopf, places it best: "Cowhide is the Python diagramming library for the individuals who need outlines now and could not care much less on the off threat that they're flawless." It's meant to work with all records types and creates graphs as SVGs, so you can scale them besides losing picture quality. Since this library is somewhat new, a component of the documentation is nevertheless in advancement. The outlines you can make are clearly essential—yet that is the goal.

Chapter 9: Introduction to Big Data

Brief History of Big Data

The origin of large volumes of data can be traced back to the 1960s and 1970s when the Third Industrial Revolution had just started to kick in, and the development of relational databases had begun along with the construction of data centers. But the concept of big data has recently taken center stage primarily since the availability of free search engines like Google and Yahoo, free online entertainment services like YouTube, and social media platforms like Facebook. In 2005, businesses started to recognize the incredible amount of user data being generated through these platforms and services, and in the same year an open-source framework called "Hadoop," was developed to gather and analyze these large data dumps available to the companies. During the same period, non-relational or distributed database called "NoSQL" started to gain popularity due to its ability to store and extract unstructured data. "Hadoop" made it possible for the companies to work with big data with high ease and at a relatively low cost.

Today with the rise of cutting-edge technology, not only humans but machines also generating data, smart device technologies like "Internet of things" (IoT) and "Internet of systems" (IoS) have skyrocketed the volume of big data. Our everyday household objects and smart devices are connected to the Internet and able to track and record our usage patterns as well as our interactions with these products and feeds all this data directly into the big data. The advent of machine learning technology has further increased the volume of data generated on a daily basis.

Big Data

You can think of it as large amounts of what appears to be random data collected on people or activities in a wide variety of circumstances. Regarding people, this will include data on where you are shopping, what websites you are visiting, where you are traveling, what products you are purchasing, and so forth. This data could also be collected on any activity related to a business or organization. It could be data collected by UPS on the habits and fuel use of their drivers, or data collected by government agencies related to criminal activity or an epidemic.

The data by itself is useless. It must be analyzed, and computers, directed by experts called data scientists, are used to find hidden patterns, associations, and trends that may exist in the data. Corporations, governments, and other large organizations are using big data to seek out information on human behavior relevant to their operations, whether that involves improving their own efficiency, delivering better services, or developing new products that their customer base will want.

Data Strategy

A big data strategy is going to involve the collection, storage, retrieval, analysis, and actionable insights. The goal of a big data strategy is to build a big data foundation that can power business intelligence.

The collection is not an issue for most businesses, but many businesses err by pre-judging the value of data that they could collect, thus not collecting as much data as they should. By now, you should recognize that data may have insights that are not readily apparent. This is the benefit of machine learning and the role that it plays in data analysis: finding hidden patterns that humans didn't know existed and - frankly – couldn't have known existed. Therefore, you need to collect data with an open mind. Any bit of information that a business can collect is potentially useful. At first glance, it might appear of no consequence, but it might be used later to draw unexpected connections that can lead to data-driven decisions later.

The second issue that needs to be addressed as part of a big data strategy is the storage and retrieval of the data. A large amount of infrastructure is needed to handle the volume and velocity of big data. This includes large storage capacity as well as powerful computers. It is difficult for organizations - especially smaller to mid-sized businesses - to build the kind of data-driven infrastructure that will enable them to handle big data effectively. Doing so could literally require hundreds of computer servers with huge banks of data storage and highly efficient data retrieval tools. In addition, a high level of cybersecurity is necessary to protect the integrity of the data.

For many businesses, these requirements mean that they are going outside the organization to have these needs met. Cloud computing has made this possible, and many of the world's largest companies are offering accessible cloud computing resources that are cost-effective, fast, and easily used to store and retrieve data on a 24/7 basis. Providers of such services include Amazon, Microsoft, and Google. These companies offer several levels of cloud services that can be tailored to fit the needs of any business size, from a one-person operation running as a home business up to a large corporation - or even a government entity. For example, Amazon's Simple Storage Service, known as Amazon S3, can be used by businesses of any size for low-cost cloud computing services.

Companies that can deploy their own big data systems can utilize open source tools to get the job done. This includes using Apache tools such as Hadoop and Spark, along with MapReduce. Careful considerations must be made. Handling the job internally will keep the data under the direct control of the company. However, this will entail having to build out or rent massive computer storage capacity that must be maintained on a 24/7 basis. That will also require a robust set of information security measures to ensure the integrity of the data and protect it from hacking, corruption, and other problems. This will also require hiring a large amount of staffing.

Even though many larger businesses might have the capacity to implement their own completely internal big data strategy, it's not clear that doing so is always the best way forward. This decision will have to be made by each organization. However, the fact that established companies like Microsoft and IBM are available for providing infrastructure and services that are already thoroughly tested and robust makes outsourcing these tasks a viable and more cost-effective option.

By utilizing third-party services, you can massively reduce the cost of implementing a big data strategy. Big data capabilities will be provided for you by established companies that have already developed massive computer infrastructure for storing, maintaining, and securing data. Third-party sources like IBM, Google, and Microsoft may also be able to assist with analysis.

Once the collection, storage, and retrieval problems are solved, systems must be put in place to make sense of the data. By itself, the data is useless. To be used for the kind of data-driven decision making that business intelligence entails, the patterns, relationships, and trends in the data must be discovered and put in a presentable form for human analysis. These steps will involve the use of data science and machine learning. This is another step where a company will have to decide whether it wants to do this internally or outsource the task.

There are several arguments that can be made for outsourcing the task for small and mid-sized businesses. Many companies like IBM have a long history of developing artificial intelligence and machine learning tools, and they are making their capabilities available as business services they are providing for clients at an effective cost. IBM has been a leader in this area since artificial intelligence first became a field, so you know you are getting high levels of reliability. This is not a recommendation for or against IBM, but rather to let you know that these types of services are available. They are also available from other large companies like Microsoft, Facebook, and Google, as well as from many newer and startup companies.

Many large corporations are doing this internally. Southwest Airlines is one example.

Either way, a part of your big data strategy is going to include developing a data science team. If you outsource your big data analysis, you may still need an internal data scientist to collaborate with third-party teams to get the most out of the data. Data scientists can be hired internally, or you can work with contract providers.

Machine learning tools for doing data analysis is also available for internal use, and you can hire data scientists to develop their own tools. Depending on your specific needs, there may be off-the-shelf software tools that you can use, and they are already proven to be reliable and effective. For this reason, you may not necessarily want to hire a team of data scientists to develop your own internal tools. Of course, if the expense can be managed and there are special needs in a company that cannot be addressed by off-the-shelf components, this can be an option to pursue as a part of your big data strategy. Another alternative is to utilize existing services from companies like IBM, Google, and Microsoft to turn your big data into insights that can be incorporated into business intelligence.

The final part of a big data strategy is engaging in data-driven decision making based on the actionable insights that have been derived from this process. This is going to become a central part of the business process, operation, and organization. This is because it is a continuous process. Machine learning is not a one-time event; it is going on constantly as large volumes of data continue to be collected. It operates in this way to be able to provide new insights often on a real-time basis after the infrastructure has been established.

In each case, the actionable insights provided by this process are going to provide a basis for business intelligence as well as provide inputs for business intelligence. Human insight and ingenuity will be brought to bear in many cases. Not every insight is going to have an equal value to the enterprise. Therefore, all the existing components of business intelligence will have to be applied in order to determine which actions can and should be taken by the organization. Issues such as cost-effectiveness are going to be central, as well as the ability of the company to carry out each action from a practical standpoint. The allocation of limited resources to different possible actions is another factor. The leaders of the business at all levels of management will have to weigh each possible action against others. This is nothing new, and businesses may have to forgo some actionable insights even if they are valuable in favor of others that are going to be impactful and be more cost-effective. The situation will have to be evaluated and re-evaluated on a consistent basis as more data continues to come in.

Not all possible actions that come out of the application of machine learning and AI to big data are going to require the input of humans. In many cases, these systems will be operating autonomously and without human input, other than a periodic review. Examples of these types of applications include cybersecurity and fraud detection in financial matters.

Harvesting Data

Raw data can be collected from disparate sources such as social media platforms, computers, cameras, other software applications, company websites, and even third-party data providers. Big data analysis inherently requires large volumes of data, the majority of which is unstructured with a limited amount of structured and semi-structured data.

Chapter 10: Neural Networks

Convolutional Neural Networks

Convolutional Neural Networks (CNNs) is one of the main categories of deep neural networks that have proven to be very effective in numerous computer science areas like object recognition, object classification, and computer vision. ConvNets have been used for many years for distinguishing faces apart, identifying objects, powering vision in self-driving cars, and robots.

A ConvNet can easily recognize countless image scenes as well as suggest relevant captions. ConvNets are also able to identify everyday objects, animals or humans, as well. Lately, Convolutional Neural Networks have also been used effectively in natural language processing problems like sentence classification.

Therefore, Convolutional Neural Networks is one of the most important tools when it comes to machine learning and deep learning tasks. LeNet was the very first Convolutional Neural Network introduced that helped significantly propel the overall field of deep learning. This very first Convolutional Neural Network was proposed by Yann LeCun back in 1988. It was primarily used for character recognition problems such as reading digits and codes.

Convolutional Neural Networks that are regularly used today for innumerable computer science tasks are very similar to this first Convolutional Neural Network proposed back in 1988.

Just like today's Convolutional Neural Networks, LeNet was used for many character recognition tasks. Just like in LeNet, the standard Convolutional Neural Networks we use today come with four main operations including convolution, ReLU non-linearity activation functions, sub-sampling or pooling and classification of their fully-connected layers.

These operations are the fundamental steps of building every Convolutional Neural Network. To move on dealing with Convolutional Neural Networks in Python, we must get deeper into these four basic functions for a better understanding of the intuition lying behind Convolutional Neural Networks.

As you know, every image can be easily represented as a matrix containing multiple values. We are going to use a conventional term channel where we are referring to a specific component of images. An image derived from a standards camera commonly has three channels including blue, red, and green. You can imagine these images as three-2D matrices that are stacked over each other. Each of these matrices also comes with certain pixel values in the specific range of 0 to 255.

On the other hand, if you have a grayscale image, you only get one channel as there are no colors present, just black and white. In our case here, we are going to consider grayscale images, so the example we are studying is just a single-2D matrix that represents a grayscale image. The value of each pixel contained in the matrix must range from 0 to 255. In this case, 0 indicates a color of black while 255 indicates a color of white.

How Convolutional Neural Networks Work?

A Convolutional Neural Network structure is normally used for various deep learning problems. As already mentioned, Convolutional Neural Networks are used for object recognition, object segmentation, detection, and computer vision due to their structure. CNNs learn directly from image data, so there is no need to perform manual feature extraction which is commonly required in regular deep neural networks.

The use of CNNs has become popular due to three main factors. The first of them is the structure of CNNs, which eliminates the need for performing manual data extraction as all data features are learned directly by the Convolutional Neural Networks. The second reason for the increasing popularity of CNNs is that they produce amazing, state-of-art object recognition results. The third reason is that CNNs can be easily retained for many new object recognition tasks to help build other deep neural networks.

A CNN can contain hundreds of layers, which each learns automatically to detect many different features of an image data. In addition, filters are commonly applied to every training image at different resolutions, so the output of every convolved image is used as the input to the following convolutional layer.

The filters can also start with very simple image features like edges and brightness, so they commonly can increase the complexity of those image features which define the object as the convolutional layers progress.

Therefore, filters are commonly applied to every training image at different resolutions as the output of every convolved image acts as the input to the following convolutional layer.

Convolutional Neural Networks can be trained on hundreds, thousands, and millions of images.

When you are working with large amounts of image data and with some very complex network structures, you should use GPUs that can significantly boost the processing time required for training a neural network model.

Once you train your Convolutional Neural Network model, you can use it in real-time applications like object recognition, pedestrian detection in ADAS or Advanced Driver Assistance Systems, and many others.

The last fully-connected layer in regular deep neural networks is called the output layer and in every classification setting, this output layer represents the overall class score.

Due to these properties, regular deep neural nets are not capable of scaling to full images. For instance, in CIFAR-10, all images are sized as 32x32x3. This means that all CIFAR-10 images gave 3 color channels and that they are 32 wide and 32 inches high. This means that a single fully-connected neural network in a first regular neural net would have 32x32x3 or 3071 weights. This is an amount that is not as manageable as those fully-connected structures are not capable of scaling to larger images.

Besides, you would want to have more similar neurons to quickly add-up more parameters. However, in this case of computer vision and other similar problems, using fully-connected neurons is wasteful as your parameters would lead to over-fitting of your model very quickly. Therefore, Convolutional Neural Networks take advantage of the fact that their inputs consist of images for solving these kinds of deep learning problems.

Due to their structure, Convolutional Neural Networks constrain the architecture of images in a much more sensible way. Unlike a regular deep neural network, the layers contained in the Convolutional Neural Network are comprised of neurons that are arranged in three dimensions including depth, height, and width. For instance, the CIFAR-10 input images are part of the input volume of all layers contained in a deep neural network and the volume comes with the dimensions of 32x32x3.

The neurons in these kinds of layers can be connected to only a small area of the layer before it, instead of all the layers being fully-connected like in regular deep neural networks. Also, the output of the final layers for CIFAR-10 would come with dimensions of 1x1x10 as the end of Convolutional Neural Networks architecture would have reduced the full image into a vector of class score arranging it just along the depth dimension.

To summarize, unlike the regular-three-layer deep neural networks, a ConvNet composes all its neurons in just three dimensions. Also, each layer contained in Convolutional Neural Network transforms the 3D input volume into a 3D output volume containing various neuron activations.

A Convolutional Neural Network contains layers that all have a simple API resulting in 3D output volume that comes with a differentiable function that may or may not contain neural network parameters.

A Convolutional Neural Network is composed of several subsamples and convolutional layers that are times followed by fully-connected or dense layers. As you already know, the input of a Convolutional Neural Network is an n x n x r image where n represents the height and width of an input image while the r is a total number of channels present. The Convolutional Neural Networks may also contain k filters known as kernels. When kernels are present, they are determined as q, which can be the same as the number of channels.

Each Convolutional Neural Network map is subsampled with max or mean pooling over p x p of a contiguous area in which p commonly ranges between 2 for small images and more than 5 for larger images. Either after or before the subsampling layer, a sigmoidal non-linearity and additive bias are applied to every feature map. After these convolutional neural layers, there may be several fully-connected layers and the structure of these fully-connected layers is the same as the structure of standard multilayer neural networks.

Stride and Padding

Secondly, after specifying the depth, you also must specify the stride that you slide over the filter. When you have a stride that is one, you must move one pixel at a time. When you have a stride that is two, you can move two pixels at a time, but this produces smaller volumes of output spatially. By default, stride value is one. However, you can have bigger strides in the case when you want to come across less overlap between your receptive fields, but, as already mentioned, this will result in having smaller feature maps as you are skipping over image locations.

In the case when you use bigger strides, but you want to maintain the same dimensionality, you must use padding that surrounds your input with zeros. You can either pad with the values on the edge or with zeros. Once you get the dimensionality of your feature map that matches your input, you can move onto adding pooling layers that padding is commonly used in Convolutional Neural Networks when you want to preserve the size of your feature maps.

If you do not use padding, your feature maps will shrink at every layer. Adding zero-padding is times very convenient when you want to pad your input volume just with zeros all around the border.

This is called as zero-padding which is a hyperparameter. By using zero-padding, you can control the size of your output volumes.

You can easily compute the spatial size of your output volume as a simple function of your input volume size, the convolution layers receptive field size, the stride you applied and the amount of zero-padding you used in your Convolutional Neural Network border.

For instance, if you have a 7x7 input and, if you use a 3x3 filter with stride 1 and pad 0, you will get a 5x5 output following the formula. If you have stride two, you will get a 3x3 output volume and so on using the formula as following in which W represents the size of your input volume, F represents the receptive field size of your convolutional neural layers, S represents the stride applied and P represents the amount of zero-padding you used.

$(W-F+2P)/S+1$

Using this formula, you can easily calculate how many neurons can fit in your Convolutional Neural Network. Consider using zero-padding whenever you can. For instance, if you have an equal input and output dimensions which are five, you can use zero-padding of one to get three receptive fields.

If you do not use zero-padding in cases like this, you will get your output volume with a spatial dimension of 3, as 3 are several neurons that can fit into your original input.

Spatial arrangement hypermeters commonly have mutual constraints. For instance, if you have an input size of 10 with no zero-padding used and with a filter size of three, it is impossible to apply stride. Therefore, you will get the set of your hyperparameter to be invalid and your Convolutional Neural Networks library will throw an exception or zero pad completely to the rest to make it fit.

Fortunately, sizing the convolutional layers properly, so that all dimensions included work using zero-padding, can make any job easier.

Parameter Sharing

You can use a parameter sharing scheme in your convolutional layers to entirely control the number of used parameters. If you denoted a single two-dimensional slice of depth as your depth slice, you can constrain the neurons contained in every depth slice to use the same bias and weights. Using parameter sharing techniques, you will get a unique collection of weights, one of every depth slice, and you will get a unique collection of weights. Therefore, you can significantly reduce the number of parameters contained in the first layer of your ConvNet. Doing this step, all neurons in every depth slice of your ConvNet will use the same parameters.

In other words, during backpropagation, every neuron contained in the volume will automatically compute the gradient for all its weights.

However, these computed gradients will add up over every depth slice, so you get to update just a single collection of weights per depth slice. In this way, all neurons contained in one depth slice will use the same weight vector. Therefore, when you forward the pass of the convolutional layers in every depth slice, it is computed as a convolution of all neurons' weights alongside the input volume. This is the reason why we refer to the collection of weights we get as a kernel or a filter, which is convolved with your input.

However, there are a few cases in which this parameter sharing assumption does not make any sense. This is commonly the case with many input images to a convolutional layer that comes with a certain centered structure, where you must learn different features depending on your image location.

For instance, when you have an input of several faces that have been centered in your image, you probably expect to get different hair-specific or eye-specific features that could be easily learned at many spatial locations. When this is the case, it is very common to just relax this parameter sharing scheme and simply use a locally-connected layer.

Matrix Multiplication

The convolution operation commonly performs those dot products between the local regions of the input and between the filters. In these cases, a common implementation technique of the convolutional layers is to take full advantage of this fact and to formulate the specific forward pass of the main convolutional layer representing it as one large matrix multiply.

Implementation of matrix multiplication is when the local areas of an input image are completely stretched out into different columns during an operation known as im2col. For instance, if you have an input of size 227x227x3 and you convolve it with a filter of size 11x11x3 at a stride of 4, you must take blocks of pixels in size 11x11x3 in the input and stretch every block into a column vector of size 363.

However, when you iterate this process in your input stride of 4, you get 55 locations along with both weight and height that lead to an output matrix of x column in which every column is a maximally stretched out receptive fields and where you have 3025 fields in total.

Each number in your input volume can be duplicated in multiple distinct columns. Also, remember that the weights of the convolutional layers are very similarly stretched out into certain rows as well. For instance, if you have 95 filters in size of 11x11x3, you will get a matrix of w row of size 96x363.

Conclusion

Python usage can be one of the best programming languages you can choose from. It's easy for beginners, but behind it is a feature, and even if you attach an advanced programmer, it's sturdy and makes it a great programming language.

There are many, and you can do the same with the Python program, and because

You can confuse it with some other programming language, and there is controversy

To do that, you can't do it in Python around you.

If you are really interested in a Programming language, python is the same as you familiar with your native language, do something really amazing without having to worry that all the code will be view. For some people, half of the fear of using a programming language is a fact. It's hard to find all the mounts and other questions. But that's the problem.

The next step is to start implementing data science into your own business and seeing what results are available with this. Data science is taking over the business world, and many companies, no matter what industry they are in, have found that this kind of process is exactly what they need to not only collect the data they have but also to clean it and perform an analysis to find the insights and predictions that are inside. When data science is used in the proper manner, and we add in some Python to help create the models and more that is needed, we are going to be able to find the best way to make business decisions that improve our standing in the industry.

There are a lot of different parts that come with data science, and being able to put them all together can really help us to do better with helping our customers, finding new products to bring to market, and more. And with the help of this guidebook, we can hopefully find the best ways to beat out the competition and see the results that will work for us. It takes some time, and a good data analysis with the right algorithms from Python, but it can be one of the best ways to make some smart and sound decisions for your business.

The process of Python data science is not an easy one, and learning how to make this work for your needs, and to put all of the parts together can make a big difference in the way that you run your business, and how much success you will see when it comes to your business growing in the future. When you are ready to learn more about working with Python data science and how to make this work for your business, make sure to check out this guidebook to get started.

The additional step is to make the best use of your new-found wisdom of Python programming; data science, data analysis, and machine learning that have resulted in the birth of the powerhouse, which is the "Silicon Valley." Businesses across the industrial spectrum with an eye on the future are gradually turning into big technology companies under the shadow of their intended business model. This has been proven with the rise of the "FinTech" industry attributed to the financial institutions and enterprises across the world.

We provide real-life examples to help you understand the nitty-gritty of the underlying concepts along with the names and descriptions of multiple tools that you can further explore and selectively implement to make sound choices for the development of a desired machine learning model. Now that you have finished reading and mastered the use of Python programming, you are all set to start developing your own Python-based machine learning model as well as performing big data analysis using all the open sources readily available and explicitly described here. You can position yourself to use your deep knowledge and understanding of all the cutting edge technologies obtained to contribute to the growth of any company and land yourself a new high paying and rewarding job!

Almost everyone will agree with the statement that big data has arrived in a big way and has taken the business world by storm. But what is the future of data analysis and will it grow? What are the technologies that will grow around it? What is the future of big data? Will it grow more? Or is the big data going to become a museum article soon? What is cognitive technology? What is the future of fast data? Let's look at the answers to these questions. We'll take a look at some predictions from the experts in the field of data analysis and big data to get a clearer picture.

The data volume will keep on growing. There is practically no question in the minds of people that we'll keep on developing a larger and larger quantity of data especially after taking into consideration the number of internet-connected devices and handheld devices is going to grow exponentially. The ways we undertake data analysis will show marked improvement in the upcoming years. Although SQL will remain the standard tool, we'll see other tools such as Spark emerging as a complementary method for the data analysis and their number will keep on growing as per reports.

More and more tools will become available for data analysis and some of them will not need the analyst. Microsoft and Salesforce have announced some combined features which will allow the non-coders to create apps for viewing the business data.

Real-time streaming insight into the big data will turn into a hallmark for the data winners moving forward. The users will be looking to use data for making informed decisions within real-time by using programs such as Spark and Kafka. The topmost strategic trend that will emerge is machine learning. Machine learning will become a mandatory element for big data preparation and predictive analysis in businesses going forward.

You can expect big data to face huge challenges as well, especially in the field of privacy of user details. The new private regulations enforced by the European Union intend to protect the personal information of the users. Various companies will have to address privacy controls and processes. It is predicted that most of the business ethics violations will be related to data in the upcoming years.

Soon you can pretty much expect all companies to have a chief data officer in place. Forrester says that this officer will rise in significance within a short period but certain kinds of businesses and generation gaps might decrease their significance in the upcoming future. Autonomous agents will continue to play a significant role and they will keep on being a huge trend as per Gartner. These agents include autonomous vehicles, smart advisers, virtual personal assistants, and robots.

The staffing required for the data analysis will keep on expanding and people from scientists to analysts to architects to the experts in the field of data management will be needed. However, a crunch in the availability of big data talent might see the large companies develop new tactics. Some large institutes predict that various organizations will use internal training to get their issues resolved. A business model having big data in the form of service can be seen on the horizon.

- PYTHON FOR DATA SCIENCE -

by

TechExp Academy

2020 © Copyright

Introduction

This book focuses on "python for data science," offering an in-depth exploration of the core utility of python programming language, specifically its application in data science. This book offers a platform to learn data science with python well and in one week with the ultimate python data science crash course including clear examples and step by step guidance.

The first thing that we need to take a look at here is the idea of why data science is so important and helpful for our needs. This would make it so much easier to go through the data and see what was there, and often the business intelligence tools were all that was needed to see what was available and what business decisions needed to be made.

However, the data that we are able to take a look at today is so much different. Unlike the traditional data systems, which was mostly structured, it is common for the data that companies collect today to be unstructured, or at the very least, semi-structured. This is going to make it more difficult to sort through and understand, and that is why the process of data science has expanded into what it is today.

Many businesses no matter what kind of industry they conduct business in; will find that working with data science is one of the best options for them. Data science is able to help them to really learn about their industry, and even gain a leg up on the competition. Many of the companies out there are going to already collect a lot of data and information about things like the competition, the industry, and their customers, and data science is going to help them to actually see what insights and information are inside of that data and use it for their advantage

With the above in mind, it is time to not only look at some of the benefits that come with data science, but we also need to take a look at some more information about what data science is all about, and how a business is able to use data science for their own needs.

The use of data science is becoming ever more prominent in many businesses and in a lot of different industries as well. But that doesn't really explain to us what data science is all about? We may want to figure out what it takes to do one of these analyses, how tools we need to make the predictions, and so much more. These are all questions that we are able to answer in this part of the guidebook to make things a bit easier.

First, we need to see what data science is all about. Data science is going to be a blend of algorithms, tools, and principles that come with machine learning. And all of these different things come together intending to discover covert patterns from the raw data. How is this going to be so much different than what we have seen statisticians doing for years now? The answer that we are going to get for this one is going to lie in the difference between predicting and explaining something.

A data analyst is usually going to be that person who will explain what is taking place by preparing the history that comes with the data. But then a data scientist is a bit different because they are not only working through some of that exploratory analysis from above, but they are going to use a lot of different algorithms that are advanced in machine learning to help them figure out how likely a certain event in the future is. A data scientist is able to take that data and look at is from a variety of angles, and hopefully, see it from a few angles that were not even explored in the past.

These are going to include the following:

The predictive causal analytics - This is the one we would want to use anytime that we have a model that can predict how likely an event will happen in the future. So, if you would like to provide someone with a loan on credit, you would want to use this kind of analytics to figure out how likely it is that the customer would make their payments in the future. We are able to build up a model that can perform these analytics based on how well that customer has made their payments in the past.

Then we can move on to a method that is known as prescriptive analytics. This is the one that we use if we want to work with a model that has the intelligence and the information that is needed to make its own verdict, and you still get the capacity to revise the model with your own dynamic criteria. This is a newer field and there are still some kinks that are being worked out with it, but it can often help us with getting advice and knowing what course of action to take out of several options that are available.

The third option that we are going to spend some time focusing on here is how we can work with machine learning to help us make some good predictions. If you are working with something like transactional data in the financial world, and you would like to build up a new model to determine the future trend that maybe there, then you will find that a few of the algorithms that work with machine learning will be the best.

This is going to be a version of supervised learning because we have to provide the system with a bunch of examples so that it can learn how to behave over time. A good example of how this one will work is when we create a model or algorithm that can detect fraud, based on historical records of purchases that ended up being fraudulent.

The last part that can be included in this is using machine learning to help us discover some new patterns in our data. If you are going into the data and you are not sure what parameters you will work with to help make some predictions, then it is time to dig in and find out what patterns are present in your batch of data that you can then employ to make some predictions that are pretty meaningful. This is going to be the same as an unsupervised learning model because the program has to go through and see what is found in the data, without any options from you or parameters, and figure out what is there that you can actually use.

Table of Contents

Chapter 1: Data Science Basics

What is Data Science?

Many definitions have been given for Data Science. We can think of Data Science as a combination of different algorithms, tools, and machine learning principles that aims to discover hidden patterns from the raw data. Data Science involves many disciples and skills such as statistics, coding, business, etc. We hear though, many buzzwords all around Data Science. Let's try to clarify them.

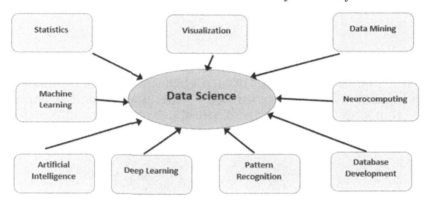

Data Analysis

Data Analysis includes the most common way of using data. Typically running analysis to understand past events and the present.

Predictive Analytics

Predictive analytics encompasses the use of techniques from machine learning, data, and statistical algorithms to help identify the likelihood of future outcomes based on data that is more historical. The goal is to go beyond what we know happened in the past to provide the best predictions and guess what will happen in the future.

Artificial Intelligence

Artificial intelligence is a technique that is used to make machines mimic any human behavior. The aim is to ensure that a machine can accurately and efficiently mimic any human behavior. Some examples of artificial intelligence machines include deep blue chess and IBM's Watson.

Machine Learning

Machine Learning can be defined as part of artificial intelligence technology driven by the hypothesis that machines are capable of learning from data by identifying patterns and making decisions with little to no human assistance

Deep Learning

Deep learning refers to a sub-branch of machine learning that involves algorithms that are motivated by the function and structure of the brain known as the artificial neural networks. This makes machines to do what is natural to humans, and that is, learn from the past. This is the system behind the idea of driverless cars.

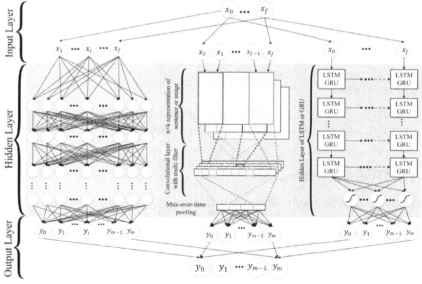

Understanding the Process of Collecting, Cleaning, Analyzing, Modeling and Visualizing Data

To follow the correct process when embarking on a Data Science project is paramount. It requires multiple iterations. Arriving at the final result requires the data scientist to refine each of the steps based on the information uncovered. Let's see what is the general process.

1. Definition of the problem or requirement

Often it is not clear what the need of the business is or we point the finger at something that is not the real problem and we seek (or try) solutions with very high efforts both in terms of costs and times.

So in this phase, the data scientist must confront those who are experiencing the problem in the business context and define it clearly and unequivocally.

2. Data collection

The data necessary to solve the problem may come from different sources such as all web channels, site, and social networks, but also company archives, CRM, marketing campaigns in progress, presence on marketplaces or offline places, results at fairs and events, up to market and competitors analysis.

3. Data processing

In this phase, errors in the data are corrected, or they are simply segmented or filtered to obtain a significant basis on which to transform them to obtain useful information.

4. Creation and selection of the model

Before thinking about automatisms and applying machine learning, or rather an automatic learning system, it is necessary to make a human and manual transition.

First of all, to have a greater knowledge of data and often also to perform a transition from big data to small data, you need to do a data mining activity. Activity that begins with identifying the relationships between the data, in order to then proceed with the real extraction of useful information also from individual actions carried out by people. Here a quote textual words from Martin Lindstrom whose book I recommend:

"While big data provides an infinite amount of impersonal information used to predict future orientations, only individual data can reveal the truth and lead to a true understanding of reality."

5. Presentation of the results

In this stage in addition to showing the results of your work, you can indulge yourself with creativity. This is a phase in which we take great responsibility, draw conclusions on the initial problem and suggest solutions or show opportunities.

Why should you Learn Python for Data Science?

The Python language is one of the best options to work with when it is time to handle the data science project that you would like to do. There are a ton of projects that you are able to handle with data science, and Python is able to take on even the more complex problems that you are trying to handle. There are many reasons why Python is going to be the best option to help you get all of this done, including:

Easy to learn. This language, even though there are a lot of complexities that come with it, is easy for a lot of beginners to learn. If you are just beginning with coding to help out with data science and you are worried about getting started, you will find that Python is the perfect one to choose from. It is designed to be as easy to work with as possible, and even those who have never done anything with coding ever will find that working with Python is not as bad as it may seem in the beginning.

Choice of data science libraries. There are a lot of special extensions and libraries available with Python that is perfect for data science and machine learning. Even though there is a lot that we are able to do when it comes to using the Python language to help out just with the standard library, there are also a number of libraries that work not only with Python, but also help us to get more done with data science at the same time.

Python community. The community that comes with Python is going to help to make this a lot easier to work with. You will be able to find a bunch of other programmers, some of whom have been using Python for a long time, who will be able to offer you advice, provide you with some helpful codes, and can get you out of the issues that you are dealing with when it comes to a data science project.

Graphics and visualization. Data Visualization is an important element for every data scientist. During the early periods of a project, you will need to perform an exploratory data analysis to identify insights into your data. Creating visualizations allows you to simplify things, particularly with a wide dimensional dataset. Towards the end of your project, you need to deliver the final result in a transparent and compelling manner that your audience can understand.

Bottom line. There are a lot of reasons why a data scientist will want to work with the Python language. There may be a few other coding languages that are out there that you are able to use, but thanks to all of the reasons that we have discussed above, and more, we are able to see that working with the Python language is one of the best options to work with that can help you create the models that you need, work with a variety of libraries that help with machine learning, and so much more.

Chapter Summary

In the world of technology, Data is defined as "information that is processed and stored by a computer". Our digital world has flooded our realities with data. From a click on a website to our smartphones tracking and recording our location every second of the day, our world is drowning in the data. From the depth of this humongous data, solutions to our problems that we have not even encountered yet could be extracted. This very process of gathering insights from a measurable set of data using mathematical equations and statistics can be defined as "data science."

Chapter 2: Setting Up the Python Environment for Data Science

Python is able to work with any kind of operating system or platform that you would like to focus on. This is good news because it ensures that we are able to go through and work with this language without changing up our system, no matter what system is the one that we are using, or the one that we will want to work with for this project.

The first thing that we need to consider here is which version of Python we would like to work with. There are a few different options here, which are going to ensure that we are going to be able to pick the version that is right for you. Python is an open-sourced language, which means that it is free to install and use, and can be modified to fit your needs. There is still a development team that works on fixing the bugs and making the updates that are necessary. This is good news for you, but it does mean that you have to figure out which version is the right one for you.

There are Python 2 and Python 3 right now, with a few updates that fit into each one. They are similar, but they will often depend on the kinds of features that you would like to use. If your computer is already providing you with a version of Python 2, then this is a good one to work with, and you don't have to update it if you do not want to in this process. It is going to be pretty similar to Python 3 and some of the codes that we will do in this guidebook.

However, most programmers are going to work with Python 3. This is the newest version of the Python language that is out there, and it is going to have some of the best features and more that you need for coding. And it is likely that at some point, the developers who work on Python 2 will stop providing support for that version, so going with the newer option is going to be the best option to go with.

You will be able to find both of these download(s) on the online platforms to work with overall, so you have to choose which one is the best for you. You can also find versions of these that will work on all of the main operating systems so it should not be too hard to add these onto whichever system you are working with.

Another thing to keep in mind here is that with the coding we are going to look at below, and with the steps that we will look at in order to install Python on all three of the major operating systems, is that we are going to get our files from www.python.org. Python is going to come with a compiler, a text editor, and an IDE, along with some other files as well. These are all important to ensure that the program is going to run properly, and if one is missing, and then you will not be able to get any of the codes that you want to work on.

There may be some other third-party sights that you are able to give a look at and then download those instead. This is an option, and depending on what you would like to do with the Python language when you are done, you may want to do this. But not all of them are going to provide you with all of the extras that are needed. You will have to go through and download each of the parts to make sure that you are not missing something.

On the other hand, when you download from the link that is provided at www.python.org, you will be able to know for a certainty that all of the files and parts that you need are there. you just need to go to the link there, click on the operating system that you are working with, and then move on to let it download on your system without any problems along the way at all.

Now that we have had some time to take a look at some of the things about installing Python on your computer, no matter which operating system you will use, it is time to take a look at the specific steps that are needed to make this happen.

Installing Python on Mac OS X

The first option that we are going to take a look at when it is time to install Python on our computers is the Mac OS X. If you are working off a computer that will rely on the Mac operating system, then you should notice early on that there will be some kind of version of Python 2 already on the computer. The exact version is something that can change based on how long you have had your computer, but we can easily check on which version by opening up our terminal app and work with the prompt below to get it done.

Python – V

This is going to show you the version you get so a number will come up. You can also choose to install Python 3 on this system if you would like, and it isn't required to uninstall the 2.X version on the computer. To check for the 3.X installation, you just need to open up the terminal app and then type in the following prompt.

Python3 – V

The default on OS X is that Python 3 is not going to be installed at all. If you want to use Python 3, you can install it using some of the installers that are on Python.org. This is a good place to go because it will install everything that you need to write and execute your codes with Python. It will have the Python shell, the IDLE development tools, and the interpreter. Unlike what happens with Python 2.X, these tools are installed as a standard application in the Applications folder.

You have to make sure that you are able to run both the Python IDLE and the Python shell on any computer that you are using. And being able to run these can be dependent on which version you have chosen for your system as well. You will need to go through and check on your system which version of Python is there before proceeding and then moves on from there to determine which IDLE and shell you need to work with.

Installing Python on a Windows System

The second option that we are going to take a look at when it is time to work with the Python language on an operating system is on a Windows computer. There are many people who rely on the Windows operating system, so it is important to know how to go through this process and how to get Python set up. This process is going to take a bit longer than the other two, mainly because Windows already has its own coding language pre-installed, so we have to take some of the extra steps that are needed to get this done.

The good news is that even though Python is going to not be the automatic option on your computer doesn't mean that it won't work great. You will find that this coding language will be fine on Windows, and there will be a lot that you are able to do with it on these systems. It is just there to give you some notice that you have to take on a few extra steps.

Now that we have that out of the way, some of the steps that you will need to follow in order to make sure that the Python language is set up and ready to go will include the following:

1. To set this up, you need to visit the official Python download page and grab the Windows installer. You can choose to do the latest version of Python 3, or go with another option. By default, the installer is going to provide you with the 32-bit version of Python, but you can

choose to switch this to the 64-bit version if you wish. The 32-bit is often best to make sure that there aren't any compatibility issues with the older packages, but you can experiment if you wish.

2. Now right-click on the installer and select "Run as Administrator." There are going to be two options to choose from. You will want to pick out "Customize Installation."

3. On the following screen, make sure all of the boxes under "Optional Features" are clicked and then click to move on.

4. When you are still under the part for Advanced Options, you need to spend a few moments figuring out where you would like the Python files to be installed. You can then click on the Install button. This will take a few minutes to finish, so have some patience and wait. Then, when the installation is done, you will be able to close out of it.

5. After this part is done, it is time for us to go through and create the PATH variable that we need for the system. We want to have it set up so that it includes the directories and all of the packages and other components that are needed later on.

The steps that we need to use to make all of this fit together will include:

1. First, we need to get the Control Panel opened up. We are going to do this by clicking on the taskbar and then typing in the Control Panel. Then we are going to click on the icon.

2. Inside of this same Control Panel, do a search for the table of Environment. Then we can click on the button to Edit the System Environment Variables. When we are on that part, we are able to click on the part that says Environment Variables.

3. Inside of the Environment Variables, we are going to find a tab for User Variables. You get two choices here of

either editing the PATH variable that you find there, or you can create a brand new one.

4. If there isn't a variable for the PATH already on the system you are using; then it is time for us to go through and create our own. You can do this by clicking on the button for New, make a name for the variable of that PATH, and then add in the directories that you would like.

Then it is time to click or close out of all the dialogs that have come up for the Control Panel and go to the next step.

1. When you reach this part, you should be able to open up the Command Prompt. This is easy to do if you go to the start menu. Then click on Windows System and then on Command Prompt. Type in the word python. This will help us to get the interpreter of Python up and running and ready to go for us.

When we have the steps above done, which will really only take a few minutes overall and won't be as bad as it may seem when you first get started, you will be able to use the Python program on your Windows system. You can then choose to open up the interpreter of Python and anything else that you need, and writing out all of the codes that you need in no time.

Installing Python on a Linux Operating System

And finally, we need to take a moment to see some of the steps that are needed to make sure that we can properly install the Python language on our Linux operating system. Linux is not quite as big of a household name as we will find with the other two operating systems that we spent some time within this guidebook. But there are so many uses for it, and it is able to provide us with a bunch of neat features and more. There are many programmers who already use this language, and many more who are likely to do it in the future, so it is a good thing to know a little bit more about when it is time to install our Python coding language on it.

If you are relying on an older version of Ubuntu or another version, then you may want to work with the dead snakes PPA, or another tool, to help you download the Python 3.6 version.

The good news here is that if you are working with other distributions of Linux, it is likely that you already have Python 3 installed on the system.

If not, you can use the distribution's package manager. And if the package of Python 3 is not recent enough, or not the right one for you, you can go through and use these same steps to help install a more recent version of Python on your computer as needed.

And there we have it! We have now gone through and downloaded Python onto the three major operating systems that are out there. This is going to make things so much easier when you are ready to handle some of the codings that you want. And if you used the codes that we described, and went through the other steps that we talked about, you will find that working with the Python language is going to be easy. You can now just open up the right files and all of the parts that you need, and it will work well for you!

Chapter 3: Seaborn Data Visualization Module in Python

Seaborn is a Matplotlib wrapper that simplifies the creation of common statistical graphs. The list of supported charts includes univariate and bivariate distribution charts, regression charts, and a number of methods for tracking categorical variables. The complete list of graphs provided by Seaborn can be found in the API reference.

Creating charts in Seaborn is as simple as calling the appropriate graphics function. Here is an example of creating a histogram, estimating kernel density and carpet texture for randomly generated data.

```
import numpy as np  # numpy used to create data from plotting
import seaborn as sns  # common form of importing seaborn

# Generate normally distributed data
data = np.random.randn(1000)

# Plot a histogram with both a rugplot and kde graph superimposed
sns.distplot(data, kde=True, rug=True)
```

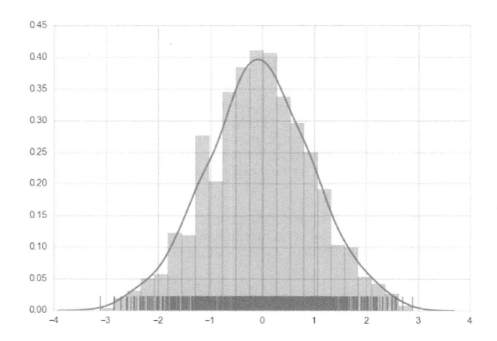

Installing Seaborn

To install the Seaborn library, you simply have to execute the following command at your command terminal:
$ pip install seaborn

Before you go and start plotting different types of plot, you need to import a few libraries. The following script does that:

import matplotlib.pyplot as plt

import seaborn as sns

plt.rcParams[«figure.figsize»] = [10,8]

tips_data = sns.load_dataset('tips')

tips_data.head()

The above script imports the Matplotlib and Seaborn libraries. Next, the default plot size is increased to 10 x 8. After that, the load_dataset() method of the Seaborn module is used to load the tips dataset. Finally, the first five records of the tips dataset have been displayed on the console. Here is the output.

Output: The tips data set contains records of the bill paid by a customer at a restaurant. The dataset contains six columns: total_bill, tip, sex, smoker, day, time, and size. You do not have to download this dataset as it comes built-in with the Seaborn library. We will be using the tips dataset to plot some of the Seaborn plots. So, without any ado, let's start plotting with Seaborn.

Seaborn's Plotting Functions

The Dist Plots

The dist plot, also known as the distributional plot, is used to plot the histogram of data for a specific column in the dataset. To plot a dist plot, you can use the distplot() function of the Seaborn library. The name of the column for which you want to plot a histogram is passed as a parameter to the distplot() function. The following script plots the dist plot for the total_bill column of the tips dataset.

Script 1:

```
plt.rcParams[«figure.figsize»] = [10,8]
sns.distplot(tips_data['total_bill'])
```

Similarly, the following script plots a dist plot for the tip column of the tips dataset.

Script 2:

```
sns.distplot(tips_data['tip'])
```

The line on top of the histogram shows the kernel density estimate for the histogram. The line can be removed by passing False as the value for the kde attribute of the distplot() function, as shown in the following example.

Script 3:

```
sns.distplot(tips_data['tip'], kde = False)
```

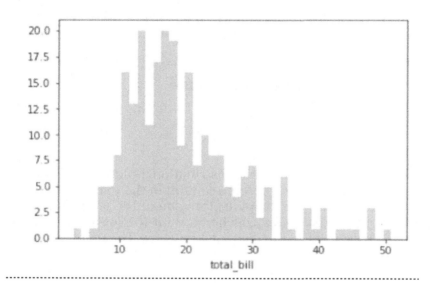

Joint Plot

The joint plot is used to plot the histogram distribution of two columns, one on the x-axis and the other on the y-axis. A scatter plot is by default drawn for the points in the two columns. To plot a joint plot, you need to call the jointplot() function. The following script plots a joint plot for the total_bill and tip columns of the tips dataset.

Script 4:

```
sns.jointplot(x='total_bill', y='tip', data=tips_data)
```

The scatter plot can be replaced by a regression line in a joint plot. To do so, you need to pass reg as the value for the kind parameter of the jointplot() function.

Script 5:

sns.jointplot(x='size', y='total_bill', data=tips_data, kind = 'reg')

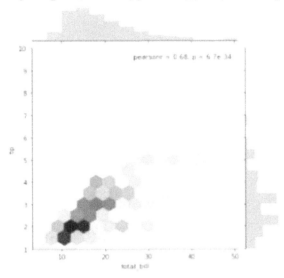

Pair Plot

The pair plot is used to plot a joint plot for all the combinations of numeric and Boolean columns in a dataset. To plot a pair plot, you need to call the pairplot() function and pass it to your dataset.

Script 6:

sns.pairplot(data=tips_data)

You can also plot multiple pair of plots per value in a categorical column. To do so, you need to pass the name of the categorical column as the value for the hue parameter. The following script plots two pair plots (one for lunch and one for dinner) for every combination of numeric or Boolean columns.

Script 7:

sns.pairplot(data=tips_data, hue = 'time')

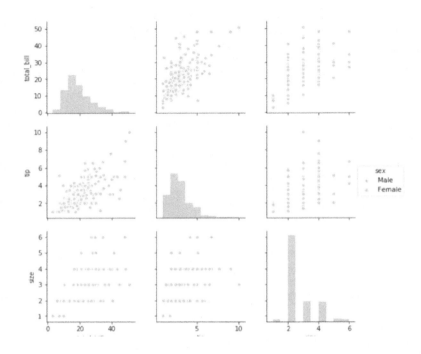

Rug Plot

The rug plot is the simplest of all the Seaborn plots. The rug plot basically plots small rectangles for all the data points in a specific column. The rugplot() function is used to plot a rug plot in Seaborn. The following script plots a rugplot() for the total_bill column of the tips dataset.

Script 8:

```
sns.rugplot(tips_data['total_bill'])
```

Output:
You can see a high concentration of rectangles between 10 and 20, which shows that the total bill amount for most of the bills is between 10 and 20.

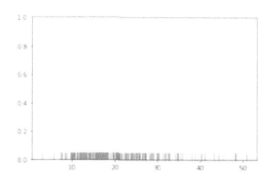

Bar Plot

The bar plot is used to capture the relationship between a categorical and numerical column. For each unique value in a categorical column, a bar is plotted, which by default, displays the mean value for the data in a numeric column specified by the bar plot.

In the following script, we first import the built-in Titanic dataset from the Seaborn library via the load_dataset() function. You can also read the CSV file named titanic_data. csv from the resources folder.

Script 9:

```
import matplotlib.pyplot as plt

import seaborn as sns

plt.rcParams[«figure.figsize»] = [8,6]

sns.set_style(«darkgrid»)

titanic_data = sns.load_dataset('titanic')

titanic_data.head()
```

Here are the first five rows of the Titanic dataset.
Next, we will call the barplot() function from the Seaborn library to plot a bar plot that displays the average age of passengers traveling in different classes of the Titanic ship.
Script 10:

```
sns.barplot(x='pclass', y='age', data=titanic_data)
```

The output shows that the average age of the passengers traveling in the first class is between 35 and 40. The average age of the passengers traveling in the second class is around 30, while the passengers traveling in the 3rd class have an average age of 25.

You can further categorize the bar plot using the hue attribute. For example, the following bar plot plots the average ages of passengers traveling in different classes and is further categorized based on their genders.

Script 11:

```
sns.barplot(x='pclass', y='age', hue ='sex', data=titanic_data)
```

The output shows that irrespective of gender, the passengers traveling in the upper classes are on average older than the passengers traveling in the lower classes.

You can also plot multiple bar plots depending upon the number of unique values in a categorical column. To do so, you need to call the catplot() function and pass the categorical column name as the value for the col attribute column. The following script plots two bar plots—one for the passengers who survived the Titanic accident and one for those who didn't survive.

Script 12:

```
sns.catplot(x=»pclass», y=»age», hue=»sex», col=»survived», data=titanic_data, kind=»bar»);
```

Output:

Role of Pandas

Filtering Rows

One of the most common tasks that you need to perform while handling the Pandas dataframe is to filter rows based on the column values.

To filter rows, first, you have to identify the indexes of the rows to filter. For those indexes, you need to pass True to the opening and closing square brackets that follow the Pandas dataframe name.

The following script returns a series of True and False. True will be returned for indexes where the Pclass column has a value of 1.

Script 2:

```
titanic_pclass1= (titanic_data.Pclass == 1)

titanic_pclass1
```

0	False
1	True
2	False
3	True
4	False
	...
886	False
887	True
888	False
889	True
890	False

Name: Pclass, Length: 891, dtype: bool

Now the titanic_pclass1 series, which contains True or False, can be passed inside the opening and closing square brackets that follow the titanic_data dataframe. The result will be the Titanic dataset, containing only those records where the Pclass column contains 1.

Script 3:

```
titanic_pclass1= (titanic_data.Pclass == 1)

titanic_pclass1_data = titanic_data[titanic_pclass1]

titanic_pclass1_data.head()
```

The comparison between the column values and filtering of rows can be made in a single line, as shown below.

Script 4:

```
titanic_pclass_data = titanic_data[titanic_data.Pclass == 1]
```

```
titanic_pclass_data.head()
```

Another commonly used operator to filter rows is the isin operator. The isin operator takes a list of values and returns only those rows where the column used for comparison contains values from the list passed to isin operator as a parameter. For instance, the following script filters those rows where age is in 20, 21, or 22.

Script 5:

```
ages = [20,21,22]
age_dataset = titan-
ic_data[titanic_data[«Age»].isin(ages)]
age_dataset.head()
```

You can filter rows in a Pandas dataframe based on multiple conditions using logical and (&) and or (|) operators. The following script returns those rows from the Pandas dataframe where passenger class is 1 and passenger age is in 20, 21, and 22.

Script 6:

```
ages = [20,21,22]
ageclass_dataset = titanic_data[titanic_data[«Age»].
isin(ages) & (titanic_data[«Pclass»] == 1) ]
ageclass_dataset.head()
```

Output:

Filtering Columns
To filter columns from a Pandas dataframe, you can use the filter() method. The list of columns that you want to filter is passed to the filter() method. The following script filters Name, Sex, and Age columns from the Titanic dataset and ignores all the other columns.

Script 7:

```
titanic_data_filter = titanic_data.filter([«Name», «Sex»,
«Age»])
```

```
titanic_data_filter.head()
```

In addition to filtering columns, you can also drop columns that you don't want in the dataset. To do so, you need to call the drop() method and pass it the list of columns that you want to drop. For instance, the following script drops the Name, Age, and Sex columns from the Titanic dataset and returns the remaining columns.

Script 8:

```
titanic_data_filter = titanic_data.drop([«Name», «Sex»,
«Age»], axis = 1)
```

```
titanic_data_filter.head()
```

Output:
Concatenating Dataframes
Often times, you need to concatenate or join multiple Pandas dataframes horizontally or vertically. Let's first see how to concatenate or join Pandas dataframes vertically. We will first create two Pandas dataframes using Titanic data. The first dataframe contains rows where the passenger class is 1, while the second dataframe contains rows where the passenger class is 2.

Script 9:

```
titanic_pclass1_data = titanic_data[titanic_data.Pclass
== 1]
```

```
print(titanic_pclass1_data.shape)
```

```
titanic_pclass2_data = titanic_data[titanic_data.Pclass
== 2]
```

```
print(titanic_pclass2_data.shape)
```

```
(216, 12)
(184, 12)
```

The output shows that both the newly created dataframes have 12 columns. It is important to mention that while concatenating data vertically, both the dataframes should have an equal number of columns.

There are two ways to concatenate datasets horizontally. You can call the append() method via the first dataframe and pass the second dataframe as a parameter to the append() method. Look at the following script.

Script 10:

```
final_data = titanic_pclass1_data.append(titanic_pclass2_data, ignore_index=True)

print(final_data.shape)
```

Output:
(400, 12)

The output now shows that the total number of rows is 400, which is the sum of the number of rows in the two dataframes that we concatenated.

The other way to concatenate two dataframes is by passing both the dataframes as parameters to the concat() method of the Pandas module. The following script shows how to do that.

Script 11:

```
final_data = pd.concat([titanic_pclass1_data, titanic_pclass2_data])

print(final_data.shape)
```

Output:

(400, 12)

To concatenate dataframes horizontally, make sure that the dataframes have an equal number of rows. You can use the concat() method to concatenate dataframes horizontally, as well. However, you will need to pass 1 as the value for the axis attribute. Furthermore, to reset dataset indexes, you need to pass True as the value for the ignore_index attribute.

Script 12:

```
df1 = final_data[:200]

print(df1.shape)

df2 = final_data[200:]

print(df2.shape)

final_data2 = pd.concat([df1, df2], axis = 1, ignore_index = True)

print(final_data2.shape)
```

Output:
(200, 12)
(200, 12)
(400, 24)
Sorting Dataframes
To sort the Pandas dataframe, you can use the sort_values() function of the Pandas dataframe. The list of columns used for sorting needs to be passed to the by attribute of the sort_values() method. The following script sorts the Titanic dataset in ascending order of the passenger's age.

Violin Plot

This type of plot is the same as the box plot, but with a violin plot, we can display all components corresponding to a data point. To create a violin plot, we call the violinplot() function. The first parameter to the function is the name of the categorical column, the second parameter is the name of the numeric column while the third column is the name of the dataset.

Let us create a violin plot that shows the distribution of age against every gender. Here is the script:

```
sns.violinplot(x='sex', y='age', data=data)
```

So that the whole script is as follows:

```
import pandas as pd
import numpy as np
import matplotlib.pyplot as plt
import seaborn as sns
import sys
sys.__stdout__ = sys.stdout
data = sns.load_dataset('titanic')
sns.violinplot(x='sex', y='age', data=data)
plt.show()
```

The script should return the following plot upon execution:

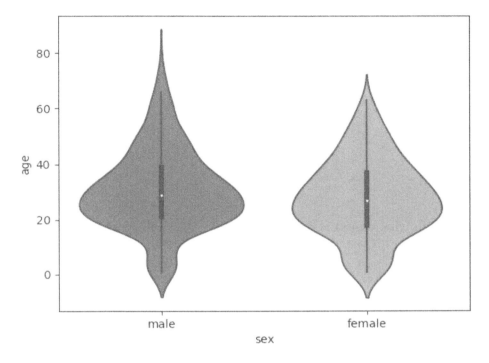

From the above plot, it is very clear that a violin plot can provide more information about the underlying data compared to a box plot. Instead of plotting the quartiles only, the violin plot shows us all components that correspond to the data. The area within the violin that is shown to be thicker means that it has a higher number of instances for the age. For the case of the violin plot for the male, it is clear that most males are aged between 20 and 40.

Line Chart

Line charts are used to display quantitative values over a continuous time period and show information as a series. A line chart is ideal for a time series, which is connected by straight-line segments.
The value is placed on the y-axis, while the x-axis is the timescale.
Uses: For smaller time periods, vertical bar charts might be the better choice.

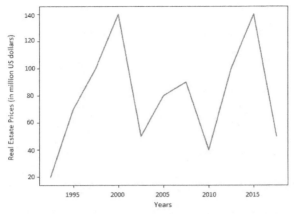

The following diagram shows a trend of real-estate prices (in million US dollars) for two decades. Line charts are well-suited for showing data trends:

Line chart for a single variable

Example:
The following diagram is a multiple variable line chart that compares the stock-closing prices for Google, Facebook, Apple, Amazon, and Microsoft. A line chart is great for comparing values and visualizing the trend of the stock. As we can see, Amazon shows the highest growth:

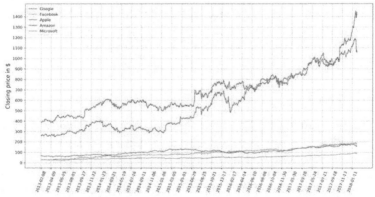

Figure: Line chart showing stock trends for the five companies

Design practices:

- Avoid too many lines per chart

- Adjust your scale so that the trend is clearly visible

Note
Design practices for plots with multiple variables. A legend should be available to describe each variable.

Bar Chart
The bar length encodes the value. There are two variants of bar charts: vertical bar charts and horizontal bar charts.

Uses:
While they are both used to compare numerical values across categories, vertical bar charts are sometimes used to show a single variable over time.
The do's and the don'ts of bar charts:
Don't confuse vertical bar charts with histograms. Bar charts compare different variables or categories, while histograms show the distribution for a single variable..
Another common mistake is to use bar charts to show central tendencies among groups or categories. Use box plots or violin plots to show statistical measures or distributions in these cases.

Examples:
The following diagram shows a vertical bar chart. Each bar shows

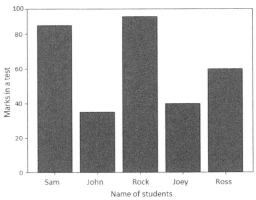

the marks out of 100 that five students obtained in a test:

Figure: Vertical bar chart using student test data
The following diagram shows a horizontal bar chart. Each bar shows the marks out of 100 that five students obtained in a test:

Figure: Horizontal bar chart using student test data
The following diagram compares movie ratings, giving two different scores. The Tomatometer is the percentage of approved critics who have given a positive review for the movie. The Audience Score is the percentage of users who have given a score of 3.5 or higher out of 5. As we can see, The Martian is the only movie with both a high Tomatometer score and an Audience Score. The Hobbit: An Unex-

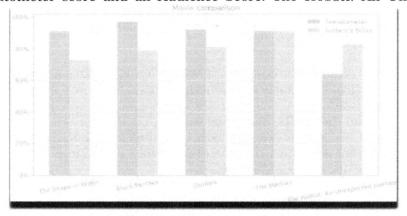

pected Journey has a relatively high Audience Score compared to the Tomatometer score, which might be due to a huge fan base:

Comparative bar chart
Design practices: The axis corresponding to the numerical variable should start at zero. Starting with another value might be misleading, as it makes a small value difference look like a big one.
Use horizontal labels, that is, as long as the number of bars is small and the chart doesn't look too cluttered.

Chapter 4: Introduction to Machine Learning Algorithms

We will take a tour of **the most popular machine learning**. There are so many algorithms that it can feel overwhelming when algorithm names are thrown around and you are expected to just know what they are and where they fit.

We want to give you two ways to think about and categorize the algorithms you may come across in the field.

- The first is a grouping of algorithms by their **learning style**.
- The second is a grouping of algorithms by their **similarity** in form or function (like grouping similar animals together).

Both approaches are useful, but we will focus in on the grouping of algorithms by similarity and go on a tour of a variety of different algorithm types.

Algorithms Grouped by Learning Style

There are different ways an algorithm can model a problem based on its interaction with the experience or environment or whatever we want to call the input data.

It is popular in machine learning and artificial intelligence textbooks to first consider the learning styles that an algorithm can adopt.

There are only a few main learning styles or learning models that an algorithm can have and we'll go through them here with a few examples of algorithms and problem types that they suit.

This taxonomy or way of organizing machine learning algorithms is useful because it forces you to think about the roles of the input data and the model preparation process and select one that is the most appropriate for your problem in order to get the best result.

Let's take a look at three different learning styles in machine learning algorithms:

Supervised Learning

Input data is called training data and has a known label or result such as spam/not-spam or a stock price at a time.

A model is prepared through a training process in which it is required to make predictions and is corrected when those predictions are wrong. The training process continues until the model achieves a desired level of accuracy on the training data. Example problems are classification and regression. Example algorithms include: Logistic Regression and the Back Propagation Neural Network.

Usupervised Learning

Input data is not labeled and does not have a known result. A model is prepared by deducing structures present in the input data. This may be to extract general rules. It may be through a mathematical process to systematically reduce redundancy, or it may be to organize data by similarity. Example problems are clustering, dimensionality reduction and association rule learning. Example algorithms include: the Apriori algorithm and K-Means.

Semi-Supervised Learning

Input data is a mixture of labeled and unlabelled examples. There is a desired prediction problem but the model must learn the structures to organize the data as well as make predictions. Example problems are classification and regression.

Example algorithms are extensions to other flexible methods that make assumptions about how to model the unlabeled data.

Overview of Machine Learning Algorithms

When crunching data to model business decisions, you are most typically using supervised and unsupervised learning methods. A hot topic at the moment is semi-supervised learning methods in areas such as image classification where there are large datasets with very few labeled examples.

Algorithms Grouped By Similarity

Algorithms are often grouped by similarity in terms of their function (how they work). For example, tree-based methods, and neural network inspired methods.
I think this is the most useful way to group algorithms and it is the approach we will use here.
This is a useful grouping method, but it is not perfect. There are still algorithms that could just as easily fit into multiple categories like Learning Vector Quantization that is both a neural network inspired method and an instance-based method. There are also categories that have the same name that describe the problem and the class of algorithm such as Regression and Clustering.
We could handle these cases by listing algorithms twice or by selecting the group that subjectively is the "BEST" fit. In this section, we list many of the popular machine learning algorithms grouped the way we think is the most intuitive. The list is not exhaustive in either the groups or the algorithms, but I think it is representative and will be useful to you to get an idea of the lay of the land.

Regression Algorithms

Regression is concerned with modeling the relationship between variables that is iteratively refined using a measure of error in the predictions made by the model.
Regression methods are a workhorse of statistics and have been co-opted into statistical machine learning. This may be confusing because we can use regression to refer to the class of problem and the class of algorithm. Really, regression is a process.
The most popular regression algorithms are:
- Ordinary Least Squares Regression (OLSR)
- Linear Regression
- Logistic Regression
- Stepwise Regression
- Multivariate Adaptive Regression Splines (MARS)

- Locally Estimated Scatterplot Smoothing (LOESS)

Instance-based Algorithms

Instance-based learning model is a decision problem with instances or examples of training data that are deemed important or required to the model.

Such methods typically build up a database of example data and compare new data to the database using a similarity measure in order to find the best match and make a prediction. For this reason, instance-based methods are also called winner-take-all methods and memory-based learning. Focus is put on the representation of the stored instances and similarity measures used between instances.

The most popular instance-based algorithms are:
- k-Nearest Neighbor (kNN)
- Learning Vector Quantization (LVQ)
- Self-Organizing Map (SOM)
- Locally Weighted Learning (LWL)
- Support Vector Machines (SVM)

Regularization Algorithms

An extension made to another method (typically regression methods) that penalizes models based on their complexity, favoring simpler models that are also better at generalizing. They are popular, powerful and generally simple modifications made to other methods.

The most popular regularization algorithms are:
- Ridge Regression
- Least Absolute Shrinkage and Selection Operator (LASSO)
- Elastic Net
- Least-Angle Regression (LARS)

Decision Tree Algorithms

Decision tree methods construct a model of decisions made based on actual values of attributes in the data. Decisions fork in tree structures until a prediction decision is made for a given record.

Decision trees are trained on data for classification and regression problems. Decision trees are often fast and accurate and a big favorite in machine learning.

The most popular decision tree algorithms are:

- Classification and Regression Tree (CART)
- Iterative Dichotomiser 3 (ID3)
- C4.5 and C5.0 (different versions of a powerful approach)
- Chi-squared Automatic Interaction Detection (CHAID)
- Decision Stump
- M5
- Conditional Decision Trees

Bayesian Algorithms

Bayesian methods are those that explicitly apply Bayes' Theorem for problems such as classification and regression. The most popular Bayesian algorithms are:

- Naive Bayes
- Gaussian Naive Bayes
- Multinomial Naive Bayes
- Averaged One-Dependence Estimators (AODE)
- Bayesian Belief Network (BBN)
- Bayesian Network (BN)

Clustering Algorithms

Clustering, like regression, describes the class of problem and the class of methods.

Clustering methods are typically organized by the modeling approaches such as centroid-based and hierarchal. All methods are concerned with using the inherent structures in the data to best organize the data into groups of maximum commonality.

The most popular clustering algorithms are:

- k-Means
- k-Medians
- Expectation Maximisation (EM)
- Hierarchical Clustering

Association Rule Learning Algorithms

Association rule learning methods extract rules that best explain observed relationships between variables in data. These rules can discover important and commercially useful associations in large multidimensional datasets that can be exploited by an organization. The most popular association rule learning algorithms are:
- Apriori algorithm
- Eclat algorithm

Artificial Neural Network Algorithms

Artificial Neural Networks are models that are inspired by the structure and/or function of biological neural networks. They are a class of pattern matching that are commonly used for regression and classification problems but are really an enormous subfield comprised of hundreds of algorithms and variations for all manner of problem types. We have separated out Deep Learning from neural networks because of the massive growth and popularity in the field. Here we are concerned with the more classical methods. The most popular artificial neural network algorithms are:
- Perceptron
- Multilayer Perceptrons (MLP)
- Back-Propagation
- Stochastic Gradient Descent
- Hopfield Network
- Radial Basis Function Network (RBFN)

Deep Learning Algorithms

Deep Learning methods are a modern update to Artificial Neural Networks that exploit abundant cheap computation. They are concerned with building much larger and more complex neural networks and, as commented on above, many methods are concerned with very large datasets of labelled analog data, such as image, text. audio, and video. The most popular deep learning algorithms are:
- Convolutional Neural Network (CNN)
- Recurrent Neural Networks (RNNs)
- Long Short-Term Memory Networks (LSTMs)

- Stacked Auto-Encoders
- Deep Boltzmann Machine (DBM)
- Deep Belief Networks (DBN)

Dimensionality Reduction Algorithms

Like clustering methods, dimensionality reduction seek and exploit the inherent structure in the data, but in this case in an unsupervised manner or order to summarize or describe data using less information. This can be useful to visualize dimensional data or to simplify data which can then be used in a supervised learning method. Many of these methods can be adapted for use in classification and regression.

- Principal Component Analysis (PCA)
- Principal Component Regression (PCR)
- Partial Least Squares Regression (PLSR)
- Sammon Mapping
- Multidimensional Scaling (MDS)
- Projection Pursuit
- Linear Discriminant Analysis (LDA)
- Mixture Discriminant Analysis (MDA)
- Quadratic Discriminant Analysis (QDA)
- Flexible Discriminant Analysis (FDA)

Ensemble Algorithms

Ensemble methods are models composed of multiple weaker models that are independently trained and whose predictions are combined in some way to make the overall prediction. Much effort is put into what types of weak learners to combine and the ways in which to combine them. This is a very powerful class of techniques and as such is very popular.

- Boosting
- Bootstrapped Aggregation (Bagging)
- AdaBoost
- Weighted Average (Blending)
- Stacked Generalization (Stacking)
- Gradient Boosting Machines (GBM)

- Gradient Boosted Regression Trees (GBRT)
- Random Forest

Chapter 5: Data Science and Machine Learning with Scikit-Learn

The Scikit-Learn library has evolved as the gold standard for the development of machine learning models with the use of Python, offering a wide variety of supervised and unsupervised ML algorithms. It is touted as one of the most user-friendly and cleanest ML libraries to date. For example, decision trees, clustering, linear and logistics regressions, and K-means. Scikit-learn uses a couple of basic Python libraries: NumPy and SciPy and adds a set of algorithms for data mining tasks including classification, regression, and clustering. It is also capable of implementing tasks like feature selection, transforming data, and ensemble methods in only a few lines.

In 2007, David Cournapeau developed the foundational code of Scikit-Learn during his participation in the "summer of code" project for Google. Scikit-learn has become one of Python's most famous open-source machine learning libraries since its launch in 2007. But it wasn't until 2010 that Scikit-Learn was released for public use. Scikit-Learn is an open-sourced and BSD licensed, data mining and data analysis tool used to develop supervise and unsupervised machine learning algorithms build on Python. Scikit-learn offers various ML algorithms such as "classification", "regression", "dimensionality reduction", and "clustering". It also offers modules for feature extraction, data processing, and model evaluation.

It is designed as an extension to the "SciPy" library, Scikit-Learn is based on "NumPy" and "Matplotlib", the most popular Python libraries. NumPy expands Python to support efficient operations on big arrays and multidimensional matrices. Matplotlib offers visualization tools and science computing modules are provided by SciPy.

For scholarly studies, Scikit-Learn is popular because it has a well-documented, easy-to-use, and flexible API. Developers are able to utilize Scikit-Learn for their experiments with various algorithms by only altering a few lines of the code. Scikit-Learn also provides a variety of training datasets, enabling developers to focus on algorithms instead of data collection and cleaning.

Many of the algorithms of Scikit-Learn are quick and scalable to all and have huge datasets. Scikit-learn is known for its reliability and automated tests are available for much of the library. Scikit-learn is extremely popular with beginners in machine learning to start implementing simple algorithms.

Prerequisites for application of Scikit-Learn library

The Scikit-Learn library is based on the SciPy (Scientific Python), which needs to be installed before you can use SciKit-Learn on your system.

Installing Scikit-Learn

The latest version of Scikit-Learn can be found on "Scikit-Learn.org" and requires "Python (version >= 3.5); NumPy (version >= 1.11.0); SciPy (version >= 0.17.0); joblib (version >= 0.11)". The plotting capabilities or functions of Scikit-learn start with "plot_" and require "Matplotlib (version >= 1.5.1)". Certain Scikit-Learn examples may need additional applications: "Scikit-Image (version >= 0.12.3), Pandas (version >= 0.18.0)".

One must make sure that binary wheels are utilized with the use of the pip files and that "NumPy" and "SciPy" have not been recompiled from source, which may occur with the use of specific OS and hardware settings (for example, "Linux on a Raspberry Pi"). Developing "NumPy" and "SciPy" from source tends to be complicated (particularly on Windows). Therefore, they need to be setup carefully, making sure the optimized execution of linear algebra routines is achievable.

Application of machine learning using Scikit-Learn library

To understand how Scikit-Learn library is used in the development of a machine learning algorithm, let us use the "Sales_Win_Loss data set from IBM's Watson repository" containing data obtained from sales campaign of a wholesale supplier of automotive parts. We will create an ML model to predict which sales campaign will be a winner and which will incur a loss.

```
#import necessary modules
import pandas as pd

#store the url in a variable
url = "https://community.watsonanalytics.com/wp-
content/uploads/2015/04/WA_Fn-UseC_-Sales-Win-Loss.csv"
```

```
#import necessary modules
import pandas as pd

#store the url in a variable
```

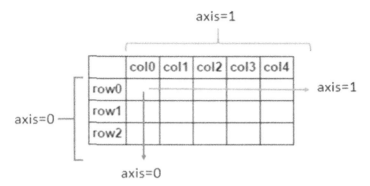

The data set can be imported with the Pandas library and explored using Pandas techniques such as "head (), tail (), and dtypes ()". The plotting techniques from "Seaborn" will be used to visualize the data. To process the data Scikit-Learn's "preprocessing.LabelEncoder ()" will be used and "train_test_split ()" to separate the data set into a training subset and testing subset.

To generate predictions from our data set, three different algorithms will be used namely, "Linear Support, Vector Classification, and K-nearest neighbors classifier". To compare the performances of these algorithms Scikit-Learn library technique, "accuracy_score", will be used. The performance score of the models can be visualized with the use of the Scikit-Learn and Yellowbrick visualization tool.

Importing the data set

To import the "Sales_Win_Loss data set from IBM's Watson repository", the first step is importing the "Pandas" module by executing the

"import pandas as pd"   command.

For new 'Pandas' users, the

"pd.read csv()"

the technique in the code mentioned above will generate a tabular data structure called "data framework", where an index for each row is contained in the first column, and the label / name for each column in the first row are the initial column names acquired from the data set. In the above code snippet, the "sales data" variable results in a table.

Now, using the "head() as Sales_data.head()" technique, the records from the data framework can be displayed to get a "feel" of the information contained in the data set.

```
# Using head() method to view the first few records of the data set
sales_data.head()
```

	Opportu nity Number	Supplies Subgrou p	Supplies Group	Region	Route To Market	Elapsed Days In Sales Stage	Opportu nity Result	Sales Stage Change Count	Total Days Identifie d Through Closing	Total Days Identifie d Through Qualifie d	Opportu nity Amount USD	Client Size By Revenue	Client Size By Employe e Count	Revenue From Client Past Two Years	Competi tor Type	Ratio Days Identifie d To Total Days	Ratio Days Validate d To Total Days	Ratio Days Qualifie d To Total Days	Deal Size Categor y
0	1641984	Exterior Accesso ries	Car Accesso ries	Northwe st	Fields Sales	76	Won	13	104	101	0	5	5	0	Unknow n	0.69636	0.11399	0.15422	1
1	1658010	Exterior Accesso ries	Car Accesso ries	Pacific	Reseller	63	Loss	2	163	163	0	3	5	0	Unknow n	0	1	0	1
2	1674737	Motorcyc le Parts	Perform ance & Non-auto	Pacific	Reseller	24	Won	7	82	82	7750	1	1	0	Unknow n	1	0	0	1
3	1675224	Shelters & RV	Perform ance & Non-auto	Midwest	Reseller	16	Loss	5	124	124	0	1	1	0	Known	1	0	0	1
4	1689785	Exterior Accesso ries	Car Accesso ries	Pacific	Reseller	69	Loss	11	91	13	69756	1	1	0	Unknow n	0	0.14113	0	4

Data Exploration

Now that we have our own copy of the data set, which has been transformed into a "Pandas" data frame, we can quickly explore the data to understand what information can be gathered from it and accordingly to planned a course of action.

In any ML project, data exploration tends to be a very critical phase. Even a fast data set exploration can offer us significant information that could be easily missed otherwise, and this information can propose significant questions that we can then attempt to answer on the basis of our project.

Some third-party Python libraries will be used here to assist us with the processing of the data so that we can efficiently use this data with the powerful algorithms of Scikit-Learn. The same "head()" technique that we used to see some initial records of the imported data set. As a matter of fact, "(head)" is effectively capable of doing much more than displaying data records and customize the "head()" technique to display only selected records with commands like "sales_data.head(n=2)".

```
# Using head() method with an argument which helps us to restrict the num-
ber of initial records that should be displayed
sales_data.head(n=2)
```

Opportunity Number	Supplies Subgroup	Supplies Group	Region	Route To Market	Elapsed Days In Sales Stage	Opportunity Result	Sales Stage Change Count	Total Days Identified Through Closing	Total Days Identified Through Qualified	Opportunity Amount USD	Client Size By Revenue	Client Size By Employee Count	Revenue From Client Past Two Years	Competitor Type	Ratio Days Identified To Total Days	Ratio Days Validated To Total Days	Ratio Days Qualified To Total Days	Deal Size Category
0 1641984	Exterior Accessories	Car Accessories	Northwest	Fields Sales	76	Won	13	104	101	0	5	5	0	Unknown	0.69636	0.11399	0.15422	1
1 1658010	Exterior Accessories	Car Accessories	Pacific	Reseller	63	Loss	2	163	163	0	3	5	0	Unknown	0	1	0	1

This command will selectively display the first 2 records of the data set. At a quick glance, it's obvious that columns such as "Supplies Group" and "Region" contain string data, while columns such as "Opportunity Result", "Opportunity Number", etc. are comprised of integer values. It can also be seen that there are unique identifiers for each record in the' Opportunity Number' column. Similarly, to display select records from the bottom of the table, the "tail() as sales_data.tail()" can be used.

```
# Using .tail() method to view the last few records from the dataframe
sales_data.tail()
```

Opportunity Number	Supplies Subgroup	Supplies Group	Region	Route To Market	Elapsed Days In Sales Stage	Opportunity Result	Sales Stage Change Count	Total Days Identified Through Closing	Total Days Identified Through Qualified	Opportunity Amount USD	Client Size By Revenue	Client Size By Employee Count	Revenue From Client Past Two Years	Competitor Type	Ratio Days Identified To Total Days	Ratio Days Validated To Total Days	Ratio Days Qualified To Total Days	Deal Size Category
78020 1E+07	Batteries & Accessories	Car Accessories	Southeast	Reseller	0	Loss	2	0	0	250000	1	1	3	Unknown	0	0	0	6
78021 1E+07	Shelters & RV	Performance & Non-auto	Southeast	Reseller	0	Won	1	0	0	180000	1	1	0	Unknown	0	0	0	5
78022 1E+07	Exterior Accessories	Car Accessories	Southeast	Reseller	0	Loss	2	0	0	90000	1	1	0	Unknown	0	0	0	4
78023 1E+07	Exterior Accessories	Car Accessories	Southeast	Fields Sales	0	Loss	2	0	0	120000	1	1	0	Unknown	1	0	0	4
78024 1E+07	Interior Accessories	Car Accessories	Mid Atlantic	Reseller	0	Loss	1	0	0	90000	1	1	0	Unknown	0	0	0	4

To view the different data types available in the data set, the Pandas technique "dtypes() as sales_data.dtypes" can be used. With this information, the data columns available in the data framework can be listed with their respective data types. We can figure out, for example, that the column "Supplies Subgroup" is an "object" data type and that the column "Client Size By Revenue" is an "integer data type". So, we have an understanding of columns that either contains integer values or string data.

```
# using the dtypes() method to display the different datatypes available
sales_data.dtypes
```

```
Opportunity Number int64
Supplies Subgroup object
Supplies Group object
Region object
Route To Market object
Elapsed Days In Sales Stage int64
Opportunity Result object
Sales Stage Change Count int64
Total Days Identified Through Closing int64
Total Days Identified Through Qualified int64
Opportunity Amount USD int64
Client Size By Revenue int64
Client Size By Employee Count int64
Revenue From Client Past Two Years int64
Competitor Type object
Ratio Days Identified To Total Days float64
Ratio Days Validated To Total Days float64
Ratio Days Qualified To Total Days float64
Deal Size Category int64
dtype: object
```

Data Visualization

At this point, we are through with basic data exploration steps, so we will not attempt to build some appealing plots to portray the information visually and discover other concealed narratives from our data set. Of all the available python libraries providing data visualization features, "Seaborn" is one of the best available options so we will be using the same. Make sure that python plots module provided by "Seaborn" has been installed on your system and ready to be used. Now follow the steps below, generate the desired plot for the data set:

Step 1 - Import the "Seaborn" module with the command "import seaborn as sns".

Step 2 - Import the "Matplotlib" module with command "import matplotlib.pyplot as plt".
Step 3 - To set the "background color" of the plot as white, use command "sns.set(style="whitegrid", color_codes=True)".

Step 4 - To set the "plot size" for all plots, use command "sns.set(rc={'figure.figsize':(11.7,8.27)})".

Step 5 – To generate a "countplot", use command "sns.countplot('Route To Market', data=sales_data, hue = 'Opportunity Result')".

Step 6 – To remove the top and bottom margins, use command "sns.despine(offset=10, trim=True)".

Step 7 – To display the plot, use command "plotplt.show()".

```python
# import the seaborn module
import seaborn as sns

# import the matplotlib module
import matplotlib.pyplot as plt

# set the background colour of the plot to white
sns.set(style="whitegrid", color_codes=True)

# setting the plot size for all plots
sns.set(rc={'figure.figsize':(11.7,8.27)})

# create a countplot
sns.countplot('Route To Market',data=sales_data,hue = 'Opportunity Result')

# Remove the top and down margin
sns.despine(offset=10, trim=True)

# display the plot
plt.show()
```

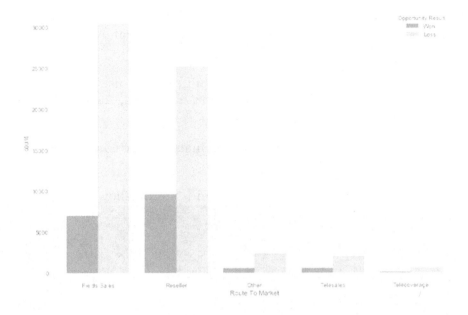

Quick recap - The "Seaborn" and "Matplotlib" modules were imported first. Then the "set()" technique was used to define the distinct characteristics for our plot, such as plot style and color. The background of the plot was defined to be white using the code snippet "sns.set(style= "whitegrid", color codes= True)". Then the plot size was defined using command "sns.set(rc= {'figure.figsize': (11.7,8.27)})" that define the size of the plot as "11.7px and 8.27px".

Next the command "sns.countplot('Route To Market',data= sales data, hue='Opportunity Result')" was used to generate the plot. The "countplot()" technique enables creation of a count plot, which can expose multiple arguments to customize the count plot according to our requirements. As part of the first "countplot()" argument, the X-axis was defined as the column "Route To Market" from the data set. The next argument concerns the source of the data set, which would be "sales_data" data framework. The third argument is the color of the bar graphs that were defined as "blue" for the column labeled "won" and "green" for the column labeled "loss".

Another options to display data is to use violin plot from Searbon.

```
# plotting the violinplot

sns.violinplot(x="Opportunity Result",y="Client Size By Revenue",
hue="Opportunity Result", data=sales_data);

plt.show()
```

A violin plot shows the distribution of data. We can see the data won "blue" and loss "green" and the client by revenue size we can see that the majority of data is for the client size "1"

Data Pre-processing

By now, you should have a clear understanding of what information is available in the data set. From the data exploration step, we established that majority of the columns in our data set are "string data", but "Scikit-Learn" can only process numerical data. Fortunately, the Scikit-Learn library offers us many ways to convert string data into numerical data, for example, "LabelEncoder()" technique. To transform categorical labels from the data set such as "won" and "loss" into numerical values, we will use the "LabelEncoder()" technique.

Let's look at the pictures below to see what we are attempting to accomplish with the "LabelEncoder()" technique. The first image contains one column labeled "color" with three records, namely, "Red",

"Green" and "Blue". By utilizing the "LabelEncoder()" technique, the record in the same "color" column can be converted to numerical values.

Let's begin the real process of conversion now. Using the "fit transform()" technique given by "LabelEncoder()", the labels in the categorical column like "Route To Market" can be encoded and converted to numerical labels comparable to those shown in the diagrams above. The function "fit transform()" requires input labels identified by the user and consequently returns encoded labels.

To know how the encoding is accomplished, let's go through an example quickly. The code instance below constitutes string data in form of a list of cities such as ["Paris", "Paris", "Tokyo", "Amsterdam"] that will be encoded into something comparable to "[2, 2, 1,3]".

Step 1 - To import the required module, use command "from sklearn import preprocessing".

Step 2 – To create the Label encoder object, use command "le = preprocessing.LabelEncoder()".

Step 3 – To convert the categorical columns into numerical values.

```
#import the necessary module
from sklearn import preprocessing

# create the LabelEncoder object
le = preprocessing.LabelEncoder()

#convert the categorical columns into numeric
encoded_value = le.fit_transform(["paris", "paris", "tokyo", "amsterdam"])

print(encoded_value)
[1 1 2 0]
```

And there you have it! We just converted our string data labels into numerical values. The first step was importing the preprocessing module that offers the "LabelEncoder()" technique. Followed by the development of an object representing the "LabelEncoder()" type. Then the "fit_transform()" function of the object was used to distinguish between distinct classes of the list ["Paris", "Paris", "Tokyo", "Amsterdam"] and output the encoded values of "[1 1 20]". Did you observe that the "LabelEncoder()" technique assigned the numerical values to the classes in alphabetical order according to the initial letter of the classes, for example "(A)msterdam" was assigned code "0", "(P)aris" was assigned code "1" and "(T)okyo" was assigned code "2".

```
print("Supplies Subgroup' : ",sales_data['Supplies Subgroup'].unique())
print("Region : ",sales_data['Region'].unique())
print("Route To Market : ",sales_data['Route To Market'].unique())
print("Opportunity Result : ",sales_data['Opportunity Result'].unique())
print("Competitor Type : ",sales_data['Competitor Type'].unique())
print("'Supplies Group : ",sales_data['Supplies Group'].unique())
```

```
Supplies Subgroup' :  ['Exterior Accessories' 'Motorcycle Parts' 'Shelters & RV'
 'Garage & Car Care' 'Batteries & Accessories' 'Performance Parts'
 'Towing & Hitches' 'Replacement Parts' 'Tires & Wheels'
 'Interior Accessories' 'Car Electronics']
Region :  ['Northwest' 'Pacific' 'Midwest' 'Southwest' 'Mid-Atlantic' 'Northeast'
 'Southeast']
Route To Market :  ['Fields Sales' 'Reseller' 'Other' 'Telesales' 'Telecoverage']
Opportunity Result :  ['Won' 'Loss']
Competitor Type :  ['Unknown' 'Known' 'None']
'Supplies Group :  ['Car Accessories' 'Performance & Non-auto' 'Tires & Wheels'
 'Car Electronics']
```

In the beginning, we loaded the preprocessing module which provides the LabelEncoder() function.

Then we created an object `le` of the type `labelEncoder()`. In the next couple of lines we used the `fit_transform()` function provided by `LabelEncoder()` and converted the categorical labels of different columns like 'Supplies Subgroup', 'Region', Route To Market' into numeric labels. In doing this, we successfully converted all the categorical (string) columns into numeric values.

Now that we have our data prepared and converted it is *almost* ready to be used for building our predictive model. But we still need to do one critical thing:

Training Set & Test

A Machine Learning algorithm needs to be trained on a set of data to learn the relationships between different features and how these features affect the target variable. For this we need to divide the entire data set into two sets. One is the training set on which we are going to train our algorithm to build a model. The other is the testing set on which we will test our model to see how accurate its predictions are.

But before doing all this splitting, let's first separate our features and target variables. As before in this tutorial, we will first run the code below, and then take a closer look at what it does:

```
# select columns other than 'Opportunity Number','Opportunity Result'cols =
[col for col in sales_data.columns if col not in ['Opportunity
Number','Opportunity Result']]

# dropping the 'Opportunity Number'and 'Opportunity Result' columns

data = sales_data[cols]

#assigning the Oppurtunity Result column as target

target = sales_data['Opportunity Result']

data.head(n=2)
```

	Supplie s Subgro up	Supplie s Group	Region	Route To Market	Elapsed Days In Sales Stage	Sales Stage Change Count	Total Days Identifie d Throug h Closing	Total Days Identifie d Throug h Qualifie d	Opportu nity Amount USD	Client Size By Revenu e	Client Size By Employ ee Count	Revenu e From Client Past Two Years	Competi tor Type	Ratio Days Identifie d To Total Days	Ratio Days Validate d To Total Days	Ratio Days Qualifie d To Total Days	Deal Size Categor y
0	2	0	3	0	76	13	104	101	0	5	5	0	2	0.69636	0.11399	0.15422	1
1	2	0	4	2	63	2	163	163	0	3							

OK, so what did we just do? First, we don't need the 'Opportunity Number' column as it is just a unique identifier for each record. Also, we want to predict the 'Opportunity Result', so it should be our 'target' rather than part of 'data'. So, in the first line of the code above, we selected only the columns which didn't match 'Opportunity Number'and 'Opportunity Result' and assigned them to a variable `cols`. Next, we created a new dataframe `data` with the columns in the list `cols`. This will serve as our feature set. Then we took the 'Opportunity Result' column from the dataframe `sales_data` and created a new dataframe `target`.

That's it! We are all set with defining our features and target into two separate dataframes. Next we will divide the dataframes `data` and `target` into training sets and testing sets. When splitting the data set we will keep 30% of the data as the test data and the remaining 70% as the training data. But keep in mind that those numbers are arbitrary and the best split will depend on the specific data you're working with. If you're not sure how to split your data, the 80/20 principle where you keep 80% of the data as training data and use the remaining 20% as test data is a decent default. However, for this tutorial, we are going to stick with our earlier decision of keeping aside 30% of the data as test data. The `train_test_split()` method in scikit-learn can be used to split the data:

```
#import the necessary module

from sklearn.model_selection import train_test_split

#split data set into train and test setsdata_train, data_test, target_train,
target_test = train_test_split(data,target, test_size = 0.30, random_state =
10)
```

With this, we have now successfully prepared a testing set and a training set. In the above code first we imported the train_test_split module. Next we used the `train_test_split()` method to divide the data into a training set (data_train,target_train) and a test set (data_test,data_train). The first argument of the `train_test_split()` method are the features that we separated out in the previous section, the second argument is the target('Opportunity Result'). The third argument 'test_size' is the percentage of the data that we want to separate out as training data . In our case it's 30% , although this can be any number. The fourth argument 'random_state' just ensures that we get reproducible results every time.

Now, we have everything ready and here comes the most important and interesting part of this tutorial: building a prediction model using the vast library of algorithms available through scikit-learn.

Building The Model

There's a `machine_learning_map` available **on scikit learn's website** that we can use as a quick reference when choosing an algorithm. It looks something like this:

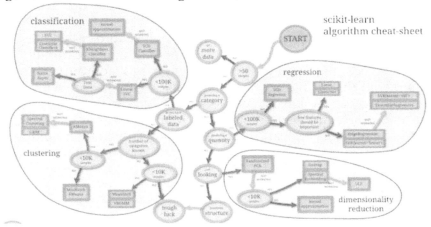

We can use this map as a cheat sheet to shortlist the algorithms that we can try out to build our prediction model. Using the checklist let's see under which category we fall:

More than 50 samples – Check

Are we predicting a category – Check

We have labeled data? (DATA WITH CLEAR NAMES LIKE OPPORTUNITY AMOUNT ETC.) – Check

Less than 100k samples – Check

Based on the checklist that we prepared above and going by the `machine_learning_map` we can try out the below mentioned algorithms.

- Naive Bayes
- Linear SVC
- K-Neighbours Classifier

The real beauty of the scikit-learn library is that it exposes high level APIs for different algorithms, making it easier for us to try out different algorithms and compare the accuracy of the models to see what works best for our data set.

Let's begin trying out the different algorithms one by one.

Naive-Bayes

Scikit-learn provides a set of classification algorithms which "naively" assumes that in a data set every pair of features are independent. This assumption is the underlying principle of **Bayes theorem**. The algorithms based on this principle are known as Naive-Bayes algorithms.

On a very high level a Naive-Bayes algorithm calculates the probability of the connection of a feature with a target variable and then it selects the feature with the highest probability. Let's try to understand this with a very simple problem statement: Will it rain today? Suppose we have a set of weather data with us that will be our feature set, and the probability of 'Rain' will be our target. Based on this feature set we can create a table to show us the number of times a particular feature/target pair occur. It would look something like this:

Weather	Rain
Partially Cloudy	No
Cloudy	Yes
Partially Cloudy	No
Partially Cloudy	Yes
Partially Cloudy	Yes
Cloudy	Yes
Total	6

In the table above the feature (column) 'Weather' contains the labels ('Partially Cloudy' and 'Cloudy') and the column 'Rain' contains the occurrence of rain coinciding with the feature 'Weather' (Yes/No). Whenever a feature lcoincides with rain, it's recorded as a 'Yes' and when the feature didn't lead to rain it is recorded as a 'No'. We can now use the data from the occurrence table to create another table known as the 'Frequency table' where we can record the number of 'Yes' and the number of 'No' answers that each feature relates to:

Frequency		
Weater	No	Yes
Partially Cloudy	2	2
Cloudy	0	2
Total	2	4

Finally, we combine the data from the 'occurrence table' and the 'frequency table' and create a 'likelihood table'. This table lists the amount of 'Yes' and 'No' for each feature and then uses this data to calculate the probability of contibution of each feature towards the occurrence of rain:

Likelihood				
Weather	No	Yes	Individual Probability	
Partially Clody	2	2	4/6	0.6666666667
Cloudy	0	2	2/6	0.3333333333
Total	2	4		
Total Probability	2/6	4/6		
	0.3333333333	0.6666666667		

Notice the 'Individual Probability' column in the table above. We had 6 occurrences of the features 'Partially Cloudy' and 'Cloudy' from the 'Occurrence table' and from the 'Likelihood table' it was clear that the feature 'Partially Cloudy' had 4 occurrences (2 for 'No' and 2 for 'yes'). When we divide the number of **OCCURRENCES** of

'No' and 'Yes' of a particular feature with the 'total' of the 'occurrence table', we get the probability of that particular feature. In our case if we need to find out that which feature has the strongest probability of contributing to the occurrence of Rain then we take the total number of 'No' of each feature and add it to their respective number of 'Yes' from the 'frequency table' and then divide the sum with the 'Total' from the óccurances table'. This gives us the probability of each of these features coinciding with rain.

The algorithm that we are going to use for our sales data is the `Gaussian Naive Bayes` and it is based on a concept similar to the weather example we just explored above, although significantly more mathematically complicated. A more detailed explanation of 'Naive-Bayes' algorithms can be found **here** for those who wish to delve deeper.

Now let's implement the Gaussian Naive Bayes or `GaussianNB` algorithm from scikit-learn to create our prediction model:

```
# import the necessary module

from sklearn.naive_bayes import GaussianNB

from sklearn.metrics import accuracy_score

#create an object of the type GaussianNB

gnb = GaussianNB()

#train the algorithm on training data and predict using the testing data

pred = gnb.fit(data_train, target_train).predict(data_test)

#print(pred.tolist())

#print the accuracy score of the model

print("Naive-Bayes accuracy : ",accuracy_score(target_test, pred, normalize
= True))
Naive-Bayes accuracy : 0.759056732741
```

Now let's take a closer look at what we just did. First, we imported the `GaussianNB` method and the `accuracy_score` method. Then we created an object `gnb` of the type `GaussianNB`. After this, we trained the algorithm on the testing data(data_train) and testing target(target_train) using the `fit()` method, and then predicted the targets in the test data using the `predict()` method. Finally we printed the score using the `accuracy_score()` method and with this we have successfully applied the `Naive-Bayes` algorithm to build a prediction model.

Now lets see how the other algorithms in our list perform as compared to the Naive-Bayes algorithm.

LinearSVC

LinearSVC or Linear Support Vector Classification is a subclass of the SVM (Support Vector Machine) class. We won't go into the intricacies of the mathematics involved in this class of algorithms, but on a very basic level LinearSVC tries to divide the data into different planes so that it can find a best possible grouping of different classes. To get a clear understanding of this concept let's imagine a data set of 'dots' and 'squares' divided into a two dimensional space along two axis, as shown in the image below:

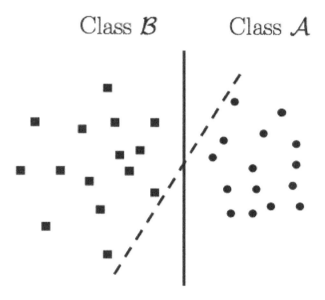

In the image above a `LinearSVC` implementation tries to divide the two-dimensional space in such a way that the two classes of data i.e the `dots` and `squares` are clearly divided. Here the two lines visually represent the various division that the `LinearSVC` tries to implement to separate out the two available classes.

A very good writeup explaining a `Support Vector Machine(SVM)` can be found **here** for those who'd like more detail, but for now, let's just dive in and get our hands dirty:

```
#import the necessary modules

from sklearn.svm import LinearSVC

from sklearn.metrics import accuracy_score

#create an object of type LinearSVC

svc_model = LinearSVC(random_state=0)

#train the algorithm on training data and predict using the testing data

pred = svc_model.fit(data_train, target_train).predict(data_test)

#print the accuracy score of the model

print("LinearSVC accuracy : ",accuracy_score(target_test, pred, normalize = True))
LinearSVC accuracy : 0.777811004785
```

Similar to what we did during the implementation of GaussianNB, we imported the required modules in the first two lines. Then we created an object `svc_model` of type LinearSVC with random_state as 'o'. Hold on! What is a "random_state" ? Simply put the `random_state` is an instruction to the built-in random number generator to shuffle the data in a specific order.

Next, we trained the LinearSVC on the training data and then predicted the target using the test data. Finally, we checked the accuracy score using the `accuracy_score()` method.

Now that we have tried out the `GaussianNB` and `LinearSVC` algorithms we will try out the last algorithm in our list and that's the `K-nearest neighbours classifier`

K-Neighbors Classifier

Compared to the previous two algorithms we've worked with, this classifier is a bit more complex.

For the purposes of this tutorial we are better off using the `KNeighborsClassifier` class provided by scikit-learn without worrying much about how the algorithm works. Now, let's implement the K-Neighbors Classifier and see how it scores:

```
#import necessary modules

from sklearn.neighbors import KNeighborsClassifier

from sklearn.metrics import accuracy_score

#create object of the lassifier

neigh = KNeighborsClassifier(n_neighbors=3)

#Train the algorithm

neigh.fit(data_train, target_train)

# predict the response

pred = neigh.predict(data_test)

# evaluate accuracy

print ("KNeighbors accuracy score : ",accuracy_score(target_test, pred))
KNeighbors accuracy score : 0.814550580998
```

The above code can be explained just like the previous implementations. First we imported the necessary modules, then we created the object `neigh` of type KNeighborsClassifier with the number of neighbors being `n_neighbors=3`. Then we used the `fit()` method to train our algorithm on the training set, then we tested the model on the test data. Finally, we printed out the accuracy score.

Now that we have implemented all the algorithms in our list, we can simply compare the scores of all the models to select the model with the highest score. But wouldn't it be nice if we had a way to visually compare the performance of the different models? We can use the `yellowbrick` library in scikit-learn, which provides methods for visually representing different scoring methods.

Performance Comparison

In the previous sections we have used the `accuracy_score()` method to measure the accuracy of the different algorithms.

Now, we will use the `ClassificationReport` class provided by the `Yellowbrick` library to give us a visual report of how our models perform.

GaussianNB

Let's start off with the `GaussianNB` model:

```python
from yellowbrick.classifier import ClassificationReport

# Instantiate the classification model and visualizer

visualizer = ClassificationReport(gnb, classes=['Won','Loss'])

visualizer.fit(data_train, target_train) # Fit the training data to the
visualizer

visualizer.score(data_test, target_test) # Evaluate the model on the test
data

g = visualizer.poof() # Draw/show/poof the data
```

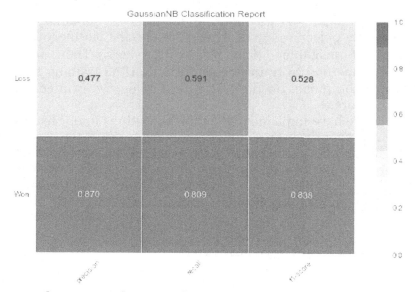

In the code above, first we import the `ClassificationReport` class provided by the `yellowbrick.classifier` module. Next, an object `visualizer` of the type `ClassificationReport` is created. Here the first argument is the `GaussianNB` object `gnb` that was created while implementing the `Naive-Bayes` algorithm in the 'Naive-Bayes' section. The second argument contains the labels

'Won' and 'Loss' from the 'Opportunity Result' column from the `sales_data` dataframe.

Next, we use the `fit()` method to train the `visualizer` object. This is followed by the `score()` method, which uses `gnb` object to carry out predictions as per the `GaussianNB` algorithm and then calculate the accuracy score of the predictions made by this algorithm. Finally, we use the `poof()` method to draw a plot of the different scores for the `GaussianNB` algorithm. Notice how the different scores are laid out against each of the labels 'Won' and 'Loss'; this enables us to visualize the scores across the different target classes.

LinearSVC

Similar to what we just did in the previous section, we can also plot the accuracy scores of the `LinearSVC` algorithm:

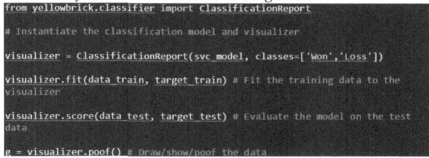

```
from yellowbrick.classifier import ClassificationReport

# Instantiate the classification model and visualizer

visualizer = ClassificationReport(svc_model, classes=['Won','Loss'])

visualizer.fit(data_train, target_train) # Fit the training data to the
visualizer

visualizer.score(data_test, target_test) # Evaluate the model on the test
data

g = visualizer.poof() # Draw/show/poof the data
```

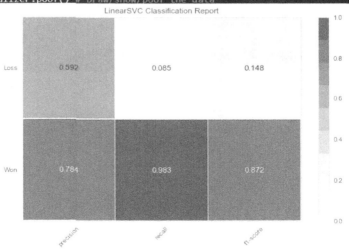

In the code above, first we imported the `ClassificationReport` class provided by the `yellowbrick.classifier` module. Next, an object `visualizer` of the type `ClassificationReport` was created. Here the first argument is the `LinearSVC` object `svc_model`, that was created while implementing the `LinearSVC` algorithm in the 'LinearSVC' section. The second argument contains the labels 'Won' and 'Loss' from the 'Opportunity Result' column from the `sales_data` dataframe.

Next, we used the `fit()` method to train the 'svc_model' object. This is followed by the `score()` method which uses the `svc_model` object to carry out predictions according to the `LinearSVC` algorithm and then calculate the accuracy score of the predictions made by this algorithm. Finally, we used the `poof()` method to draw a plot of the different scores for the `LinearSVC` algorithm.

KNeighborsClassifier

Now, let's do the same thing for the K-Neighbors Classifier scores.

```
from yellowbrick.classifier import ClassificationReport

# Instantiate the classification model and visualizer

visualizer = ClassificationReport(neigh, classes=['Won','Loss'])

visualizer.fit(data_train, target_train) # Fit the training data to the
visualizer

visualizer.score(data_test, target_test) # Evaluate the model on the test
data

g = visualizer.poof() # Draw/show/poof the data
```

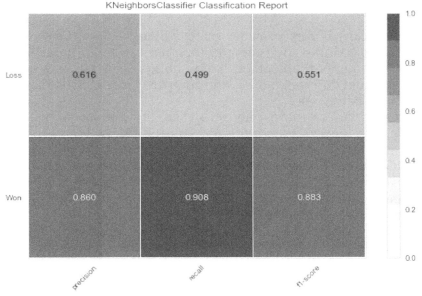

KNeighborsClassifier Classification Report

Once again, we first import the ClassificationReport class provided by the yellowbrick.classifier module. Next, an object visualizer of the type ClassificationReport is created. Here the first argument is the KNeighborsClassifier object neigh, that was created while implementing the KNeighborsClassifier algorithm in the 'KNeighborsClassifier' section. The second argument contains the labels 'Won' and 'Loss' from the 'Opportunity Result' column from the sales_data dataframe.

Next, we use the fit() method to train the 'neigh' object. This is followed by the score() method which uses the neigh object to carry out predictions according to the KNeighborsClassifier algorithm and then calculate the accuracy score of the predictions made by this algorithm. Finally we use the poof() method to draw a plot of the different scores for the KNeighborsClassifier algorithm.

Now that we've visualized the results, it's much easier for us to compare the scores and choose the algorithm that's going to work best for our needs.

Conclusion

The scikit-learn library provides many different algorithms which can be imported into the code and then used to build models just like we would import any other Python library. This makes it easier to quickly build different models and compare these models to select the highest scoring one.

In this tutorial, we have only scratched the surface of what is possible with the scikit-learn library. To use this Machine Learning library to the fullest, there are many resources available on the **official page of scikit-learn** with detailed documentation that you can dive into. The quick start guide for scikit-learn can be found **here**, and that's a good entry point for beginners who have just started exploring the world of Machine Learning.

But to really appreciate the true power of the scikit-learn library, what you really need to do is start using it on different open data sets and building predictive models using these data sets. Sources for open data sets include **Kaggle** and **Data.world**. Both contain many interesting data sets on which one can practice building predictive models by using the algorithms provided by the scikit-learn library.

Chapter 6: Data Science and Cloud

Data science is a mixture of many concepts. To become a data scientist, it is important to have some programming skills. Even though you might not know all the programming concepts related to infrastructure but having basic skills in computer science concepts is a must. With the ever-expanding advanced analytics, Data Science continues to spread its wings in different directions. This requires collaborative solutions like predictive analysis and recommendation systems. Collaboration solutions include research and notebook tools integrated with code source control. Data science is also related to the cloud. The information is also stored in the cloud. So, this lesson will enlighten you with some facts about the "data in the Cloud." So let us understand what cloud means and how the data is stored and how it works.

What is the Cloud?

Cloud can be described as a global server network, each having different unique functions. Understanding networks is required to study the cloud. Networks can be simple or complex clusters of information or data.

Network

Networks can have a simple or small group of computers connected or large groups of computers connected. The largest network is the Internet. The small groups can be home local networks like Wi-Fi, and Local Area Network that is limited to certain computers or locality. There are shared networks such as media, web pages, app servers, data storage, and printers, and scanners. Networks have nodes, where a computer is referred to as a node. The communication between these computers is established by using protocols. Protocols are the intermediary rules set for a computer. Protocols like HTTP, TCP, and IP are used on a large scale. All the information is stored on the computer, but it becomes difficult to search for information on the computer every time. Such information is usually stored in a data Centre. Data Centre is designed in such a way that it is equipped with support security and protection for the data. Since the cost of computers and storage has decreased substantially, multiple organizations opt to make use of multiple computers that work together that one wants to scale. This differs from other scaling solutions like buying other computing devices. The intent behind this is to keep the work going continuously even if a computer fails; the other will continue the operation. There is a need to scale some cloud applications, as well. Having a broad look at some computing applications like YouTube, Netflix, and Facebook that requires some scaling. We rarely experience such applications failing, as they have set up their systems on the cloud. There is a network cluster in the cloud, where many computers are connected to the same networks and accomplish similar tasks. You can call it a single source of information or a single computer that manages everything to improve performance, scalability, and availability.

Data Science in the Cloud

The whole process of Data Science takes place in the local machine, i.e., a computer or laptop provided to the data scientist. The computer or laptop has inbuilt programming languages and a few more prerequisites installed. This can include common programming languages and some algorithms. The data scientist later has to install relevant software and development packages as per his/her project. Development packages can be installed using managers such as Anaconda or similar managers. You can opt for installing them manually too. Once you install and enter into the development environment, then your first step, i.e., the workflow starts where your companion is only data. It is not mandatory to carry out the task related to Data Science or Big data on different development machines. Check out the reasons behind this:

- The processing time required to carry out tasks on the development environment fails due to processing power failure.
- Presence of large data sets that cannot be contained in the development environment's system memory.
- Deliverables must be arrayed into a production environment and incorporated as a component in a large application.
- It is advised to use a machine that is fast and powerful.

Data scientist explores many options when they face such issues; they make use of on-premise machines or virtual machines that run on the cloud. Using virtual machines and auto-scaling clusters has various benefits, such as they can span up and discard it anytime in case it is required. Virtual machines are customized in a way that will fulfill one's computing power and storage needs. Deployment of the information in a production environment to push it in a large data pipeline may have certain challenges. These challenges are to be understood and analyzed by the data scientist. This can be understood by having a gist of software architectures and quality attributes.

Software Architecture and Quality Attributes

A cloud-based software system is developed by Software Architects. Such systems may be product or service that depends on the computing system. If you are building software, the main task includes the selection of the right programming language that is to be programmed. The purpose of the system can be questioned; hence, it needs to be considered. Developing and working with software architecture must be done by a highly skilled person. Most of organizations have started implementing effective and reliable cloud environment using cloud computing. These cloud environments are deployed over to various servers, storage, and networking resources. This is used in abundance due to its less cost and high ROI.

The main benefit to data scientists or their teams is that they are using the big space in the cloud to explore more data and create important use cases. You can release a feature and have it tested the next second and check whether it adds value, or it is not useful to carry forward. All this immediate action is possible due to cloud computing.

Sharing Big Data in the Cloud

The role of Big Data is also vital while dealing with the cloud as it makes it easier to track and analyze insights. Once this is established, big data creates great value for users.

The traditional way was to process wired data. It became difficult for the team to share their information with this technique. The usual problems included transferring large amounts of data and collaboration of the same. This is where cloud computing started sowing its seed in the competitive world. All these problems were eliminated due to cloud computing, and gradually, teams were able to work together from different locations and overseas as well. Therefore, cloud computing is very vital in both Data Science as well as Big data. Most of organizations make use of the cloud. To illustrate, a few companies that use the cloud are Swiggy, Uber, Airbnb, etc. They use cloud computing for sharing information and data.

Cloud and Big data Governance

Working with the cloud is a great experience as it reduces resource cost, time, and manual efforts. But the question arises that how organizations deal with security, compliance, governance? Regulation of the same is a challenge for most companies. Not limited to Big data problems but working with the cloud also has its issues related to privacy and security. Hence, it is required to develop a strong governance policy in your cloud solutions. To ensure that your cloud solutions are reliable, robust, and governable, you must keep it as an open architecture.

Need for Data Cloud Tools to Deliver High Value of Data

Demand for a data scientist in this era is increasing rapidly. They are responsible for helping big and small organizations to develop useful information from the data or data set that is provided. Large organizations carry massive data that needs to analyze continuously. As per recent reports, almost 80% of the unstructured data received by the organizations are in the form of social media, emails, i.e., Outlook, Gmail, etc., videos, images, etc. With the rapid growth of cloud computing, data scientists deal with various new workloads that come from IoT, AI, Blockchain, Analytics, etc. Pipeline. Working with all these new workloads requires a stable, efficient, and centralized platform across all teams. With all this, there is a need for managing and recording new data as well as legacy documents. Once a data scientist is given a task, and he/she has the dataset to work on, he/she must possess the right skills to analyze the ever-increasing volumes through cloud technologies. They need to convert the data into useful insights that would be responsible for uplifting the business. The data scientist has to build an algorithm and code the program. They mostly utilize 80% of their time to gathering information, creating and modifying data, cleaning if required, and organizing data. Rest 20% is utilized for analyzing the data with effective programming. This calls for the requirement of having specific cloud tools to help the data scientist to reduce their time searching for appropriate information. Organizations should make available new cloud services and cloud tools to their respective data scientists so that they can organize massive data quickly. Therefore, cloud tools are very important for a data scientist to analyze large amounts of data at a shorter period. It will save the company's time and help build strong and robust data Models.

Chapter 7: Examples of Applications of Data Science

Security

There are several cities throughout the world that are working on predictive analysis so that they can predict the areas of the town where there is more likely to be a big surge for the crime that is there. This is done with the help of some data from the past and even data on the geography of the area.

This is actually something that a few cities in America have been able to use, including Chicago. Although we can imagine that it is impossible to use this to catch every crime that is out there, the data that is available from using this is going to make it easier for police officers to be present in the right areas at the right times to help reduce the rates of crime in some of those areas. And in the future, you will find that when we use data analysis in this kind of manner in the big cities has helped to make these cities and these areas a lot safer, and the risks would not have to put their lives at risk as much as before.

Transportation

The world of transportation is able to work with data analysis, as well. A few years ago, when plans were being made at the London Olympics, there was a need during this event to handle more than 18 million journeys that were made by fans into the city of London. Moreover, it was something that we were able to sort out well.

How was this feat achieved for all of these people? The train operators and the TFL operators worked with data analytics to make sure that all those journeys went as smoothly as possible. These groups were able to go through and input data from the events that happened around that time and then used this as a way to forecast how many people would travel to it. This plan went so well that all of the spectators and the athletes could be moved to and from the right places in a timely manner the whole event.

Risk and Fraud Detection

This was one of the original uses of data analysis and was often used in the field of finance. There are many organizations that had a bad experience with debt, and they were ready to make some changes to this. Because they had a hold on the data that was collected each time that the customer came in for a loan, they were able to work with this process in order to not lose as much money in the process.

This allowed the banks and other financial institutions to dive and conquer some of the data from the profiles they could use from those customers. When the bank or financial institution is able to utilize their customers they are working with, the costs that had come up recently, and some of the other information that is important for these tools, they will make some better decisions about who to loan out money to, reducing their risks overall. This helps them to offer better rates to their customers.

In addition to helping these financial institutions make sure that they can hand out loans to customers who are more likely to pay them back, you will find that this can be used in order to help cut down on the risks of fraud as well. This can cost the bank billions of dollars a year and can be expensive to work with. When the bank can use all of the data that they have for helping discover transactions that are fraudulent and making it easier for their customers to keep money in their account, and make sure that the bank is not going to lose money in the process as well.

Logistics of Deliveries

There are no limitations when it comes to what we are able to do with our data analysis, and we will find that it works well when it comes to logistics and deliveries. There are several companies that focus on logistics, which will work with this data analysis, including UPS, FedEx, and DHL. They will use data in order to improve how efficient their operations are all about.

From applications of analytics of the data, it is possible for these companies who use it to find the best and most efficient routes to use when shipping items, the ones that will ensure the items will be delivered on time, and so much more. This helps the item to get things through in no time, and keeps costs down to a minimum as well. Along with this, the information that the companies are able to gather through their GPS can give them more opportunities in the future to use data science and data analytics.

Customer Interactions

Many businesses are going to work with the applications of data analytics in order to have better interactions with their customers. Companies can do a lot about their customers, often with some customer surveys. For example, many insurance companies are going to use this by sending out customer surveys after they interact with their handler. The insurance company is then able to use which of their services are good, that the customers like, and which ones they would like to work on to see some improvements.

There are many demographics that a business is able to work with and it is possible that these are going to need many diverse methods of communication, including email, phone, websites, and in-person interactions. Taking some of the analysis that they can get with the demographics of their customers and the feedback that comes in, it will ensure that these insurance companies can offer the right products to these customers, and it depends one hundred percent on the proven insights and customer behavior as well.

City Planning

One of the big mistakes that is being made in many places is that analytics, especially the steps that we are talking about in this guidebook, is not something that is being used and considered when it comes to city planning. Web traffic and marketing are actually the things that are being used instead of the creation of buildings and spaces. This is going to cause many of the issues that are going to come up when we talk about the power over our data is because there are some influences over building zoning and creating new things along the way in the city.

Models that have been built well are going to help maximize the accessibility of specific services and areas while ensuring that there is not the risk of overloading significant elements of the infrastructure in the city at the same time. This helps to make sure there is a level of efficiency as everyone, as much as possible, is able to get what they want without doing too much to the city and causing harm in that manner.

We will usually see buildings that are not put in the right spots or businesses that are moved where they do not belong. How often have you seen a building that was on a spot that looked like it was suitable and good for the need, but which had a lot of negative impact on other places around it? This is because these potential issues were not part of the consideration during the planning period. Applications of data analytics, and some modeling, helps us to make things easier because we will know what would happen if we put that building or another item on that spot that you want to choose.

Healthcare

The healthcare industry has been able to see many benefits from data analysis. There are many methods, but we are going to look at one of the main challenges that hospitals are going to face. Moreover, this is that they need to cope with cost pressures when they want to treat as many patients as possible while still getting high-quality care to the patients. This makes the doctors and other staff fall behind in some of their work on occasion, and it is hard to keep up with the demand.

You will find that the data we can use here has raised so much, and it allows the hospital to optimize and then track the treatment of their patient. It is also a good way to track the patient flow and how the different equipment in the hospital is being used. In fact, this is so powerful that it is estimated that using this data analytics could provide a 1 percent efficiency gain, and could result in more than $63 billion in worldwide healthcare services. Think of what that could mean to you and those around you.

Doctors are going to work with data analysis in order to provide them with a way to help their patients a bit more. They can use this to make some diagnoses and understand what is going on with their patients in a timely and more efficient manner. This can allow doctors to provide their customers with a better experience and better care while ensuring that they can keep up with everything they need to do.

Travel

Data analytics and some of their applications are a good way to help optimize the buying experience for a traveler. This can be true through a variety of options, including data analysis of mobile sources, websites, or social media. The reason for this is because the desires and the preferences of the customer can be obtained from all of these sources, which makes companies start to sell out their products thanks to the correlation of all the recent browsing on the site and any of the currency sells to help purchase conversions. They are able to utilize all of this to offer some customized packages and offers. The applications of data analytics can also help to deliver some personalized travel recommendations, and it often depends on the outcome that the company is able to get from their data on social media.

Travel can benefit other ways when it comes to working with the data analysis. When hotels are trying to fill up, they can work with data analysis to figure out which advertisements they would like to offer to their customers. Moreover, they may try to utilize this to help figure out which nights, and which customers, will fill up or show up. Pretty much all of the different parts of the travel world can benefit when it comes to working with data analysis.

Digital Advertising

Outside of just using it to help with some searching, there is another area where we are able to see data analytics happen regularly, and this is digital advertisements. From some of the banners that are found on several websites to the digital billboards that you may be used to seeing in some of the bigger and larger cities, but all of these will be controlled thanks to the algorithms of our data along the way.

This is a good reason why digital advertisements are more likely to get a higher CTR than the conventional methods that advertisers used to rely on a lot more. The targets are going to work more on the past behaviors of the users, and this can make for some good predictions in the future.

The importance that we see with the applications of data analytics is not something that we can overemphasize because it is going to be used in pretty much any and all of the areas of our life to ensure we have things go a bit easier than before. It is easier to see now, more than ever, how having data is such an important thing because it helps us to make some of the best decisions without any issues. However, if we don't have that data or we are not able to get through it because it is a mess and doo many points to look at, then our decisions are going to be based on something else. Data analysis ensures that our decisions are well thought out, that they make sense, and that they will work for our needs.

You may also find that when we inefficiently handle our data, it could lead to a number of problems. For example, it could lead to some of the departments that are found in a larger company so that we have a better idea of how we can use the data and the insights that we are able to find in the process, which could make it so that the data you have is not able to be used to its full potential. Moreover, if this gets too bad, then it is possible that the data will not serve any purpose at all.

However, you will find that as data is more accessible and available than ever before, and therefore more people, it is no longer just something that the data analysts and the data scientists are able to handle and no one else. Proper use of this data is important, but everyone is able to go out there and find the data they want. Moreover, this trend is likely to continue long into the future as well.

- PYTHON MACHINE LEARNING -

by

TechExp Academy

2020 © Copyright

Introduction

For all that we know about Machine Learning, the truth is that we are nowhere close to realizing the true potential of these studies. Machine Learning is currently one of the hottest topics in computer science. If you are a data analyst, this is a field you should focus all your energy on because the prospects are incredible. You are looking at a future where interaction with machines will form the base of our being.

In this installation, our purpose was to address Python Machine Learning from the perspective of an expert. The assumption is that you have gone through the earlier books in the series that introduced you to Machine Learning, Python, libraries, and other important features that form the foundation of your knowledge in Machine Learning. With this in mind, we barely touched on the introductory concepts, unless necessary.

Even at an expert level, it is always important to remind yourself of the important issues that we must look at in Machine Learning. Algorithms are the backbone of almost everything that you will do in Machine Learning. Because of this reason, we introduced a brief section where you can remind yourself of the important algorithms and other elements that help you progress your knowledge of Machine Learning.

Machine Learning is as much about programming as it is about probability and statistics. There are many statistical approaches that we will use in Machine Learning to help us arrive at optimal solutions from time to time. It is therefore important that you remind yourself about some of the necessary probability theories and how they affect outcomes in each scenario.

In our studies of Machine Learning one concept that stands out is that Machine Learning involves uncertainty. This is one of the differences between Machine Learning and programming. In programming, you write code that must be executed as it is written. The code derives a predetermined output

based on the instructions given. However, in Machine Learning, this is not a luxury we enjoy.

Once you build the model, you train and test it and eventually deploy the model. Since these models are built to interact with humans, you can expect variances in the type of interaction that you experience at every level. Some input parameters might be correct, while others might not. When you build your model, you must consider these factors, or your model will cease to perform as expected.

The math element of Machine Learning is another area of study that we have to look at. Many mathematical computations are involved in Machine Learning for the models to deliver the output we need. To support this cause, we must learn how to perform specific operations on data based on unique instructions.

As you work with different sets of data, there is always the possibility that you will come across massive datasets. This is normal because as our Machine Learning models interact with different users, they keep learning and build their knowledge. The challenge of using massive datasets is that you must learn how to break down the data into small units that your system can handle and process without any challenges. In this case, you are trying to avoid overworking your learning model.

Most basic computers will crash when they have to handle massive data. However, this should not be a problem when you learn how to fragment your datasets and perform computational operations on them.

At the beginning of this book, we mentioned that we will introduce hands-on approaches to using Machine Learning in daily applications. In light of this assertion, we looked at some practical methods of using Machine Learning.

We have taken a careful step-by-step approach to ensure that you can learn along the way, and more importantly, tried to explain each process to help you understand the operations you perform and why.

Eventually, when you build a Machine Learning model, the aim is to integrate it into some of the applications that people use daily. With this in mind, you must learn how to build a simple solution that addresses this challenge. We used simple explanations to help you understand this, and hopefully, as you keep working on different Machine Learning models, you can learn by building more complex models as your needs permit.

There are many concepts in Machine Learning that you will learn or come across over time. You must reckon the fact that this is a never-ending learning process as long as your model interacts with data. Over time, you will encounter greater datasets than those you are used to working with. In such a scenario, learning how to handle them will help you achieve your results faster, and without struggling.

Table of Contents

The idea of artificial intelligence was also portrayed in the famous science fiction movie 2001: A Space Odyssey.

This was a more realistic portrayal, the computer was not human-like in form, but it clearly had a mind. One of the key themes in the film is the fact that the system has faults, and can make bad decisions as a result. This is probably a warning we should take to heart, but we probably won't. The system in the movie brags that it never makes mistakes, but it clearly does. Unfortunately, people today place too much faith in computers. Machine learning is great, but don't worship it.

Another aspect of the HAL 9000 computer portrayed in the film was actually realistic in the sense of what artificially intelligent computer systems can actually do in the real world, and what they might actually be used for. The system in the movie detects a coming fault in a communications device. In the movie, it turns out to be in error, but the point is many machine learning systems are being developed that will hopefully detect faults in electronics and machines before humans become aware of them.

We don't want to get too carried away with science fiction stories, but the analogies and interpretations of what is possible are interesting to consider.

It turns out that hype won the day, and after a few decades of pursuing the holy grail of artificial intelligence, in effect creating a computer system that worked like a human brain, computer scientists began to change direction. They turned more attention to machine learning, which is something that can be thought of as using human-like intelligence but applied to a very narrowly tailored task. Something that was missing, however, was data.

How Machine Learning Works

Machine learning begins with a human guide in the form of a data scientist. A problem has to be identified that a machine learning algorithm can be used to solve. In order for the process to work, there must be a significant amount of data that can be used to train the system before it is deployed in the real world. There are two basic ways that machine learning can proceed. They are called supervised and unsupervised learning.

In the training phase, the data scientist will select appropriate training data, and expose the system to the data. As the system is exposed to data, it will modify itself in order to become more accurate. This phase is a crucial part of a development with the system, and the data scientist must choose the right data set for the problem at hand. In other words, the impression that human beings are not involved at all is completely false. Human beings are involved in choosing and framing the problem, and in choosing the right training data. The human being is also involved in the process of evaluating the performance of the system. As we will see, there are tradeoffs that must be made that have large implications for the viability of the system. If the data scientists that are working on the problem are not careful and correct, to the best of their abilities, in interpreting the results produced by the system, a dysfunctional system can be deployed that will err when attempting to do its job.

The training phase may have several iterations, depending on how the results turn out.

Once the system is deployed, it will more or less operate in an autonomous fashion, but data scientists that are involved in this work should be continually evaluating the performance of the system. At some point, the system may need to be replaced or it may need to be subjected to more training if it is not producing the kinds of results that are accepted.

Why Use Python for AI and Machine Learning?

There are a lot of different coding languages that you can learn about and use. So, if there are a lot of choices out there, why would you want to go with the Python coding language in the first place? Many people, both experts, and beginners all choose to go with Python because it is easy to learn, easy to read, and it is capable of creating large, challenging codes that you might want to write. There are a lot of different reasons that you would want to work with this coding language, and these include:

Simple and consistent

You will find that Python is a programming language that is simple and quite easy to read, even if you are a beginner. When it is compared to some of the other coding languages, it is one of the most readable languages. Since this is a natural coding language to go with, many beginners like that they can catch on so quickly and that they will understand what they are doing in no time.

An extensive selection of libraries and frameworks

Once you start to get familiar with Python, you will notice that it comes with an extensive library. This is good news for beginners because the library is what contains all the functions, codes, and other things that you need to make the language work for you. This library will help make sure that you can do some useful stuff when trying to make your code.

Platform independence

You can work with the Python language no matter which platform you would like to use it on. Linux is the operating system that a lot of people will choose to go with, but you can still work with Python even if you are on a Windows or Mac computer. This is excellent news because it means that you can use Python without having to make any significant changes to your current setup.

Great community and popularity

Whether you have worked with coding language in the past or not, it is nice to know that there is a large community of Python users that will help you out if you ever get stuck. Any time that you need some ideas like a new project or if you have a question, or if you want to learn something new, there is a big group to provide you with the information that you need to help you get started.

Chapter 2: The 7 Steps of Machine Learning Process

The models that are based on labeled training data sets are termed supervised Machine Learning models. When children go to school, their teachers and professors give them some feedback about their progress. In the same way, a supervised Machine Learning model allows the engineer to provide some feedback to the machine.

Let us take an example of an input [red, round]. Here, both the child and machine will understand that any object which is round and red is an apple. Let us now place a cricket ball in front of either the machine or the child. You can feed the machine with the response negative 1 or 0 depending on whether the prediction is wrong or right. You can always add more attributes if necessary. This is the only way that a machine will learn. It is also for this reason that if you use a large-high-quality data set and spend more time training the machine, the machine will give you better and more accurate results.

Before we proceed further, you must understand the difference between the concepts of Machine Learning, artificial intelligence, and deep learning. Most people use these concepts interchangeably, but it is important to know that they are not the same.

Machine Learning, Artificial Intelligence and Deep Learning

The diagram below will give you an idea of how these terms relate.

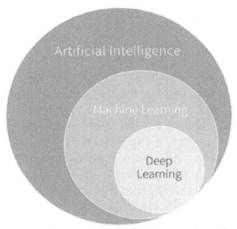

An illustration to understand the relationship between Machine Learning, Artificial Intelligence, and Deep Learning

Artificial intelligence is a technique that is used to make machines mimic any human behavior. The aim is to ensure that a machine can accurately and efficiently mimic any human behavior. Some examples of artificial intelligence machines include deep blue chess and IBM's Watson.

Machine Learning, as defined above, is the use of statistical and mathematical models to help machines learned mimic human behavior. This is done using past data.

Deep learning is a subset of Machine Learning, and it refers to some of the functions and algorithms that an engineer uses to help a machine to train itself. The machine can learn to take the correct option to derive an output. Neural networks and natural language processing are a part of the deep learning ecosystem.

Typical Objectives of Machine Learning System

The systems of Machine Learning usually are set up with one of the following objectives:
- Predict a category
- Predict a quantity

- Anomaly Detector Systems
- Clustering Systems

Predict a category

The model of Machine Learning helps analyze the input data and then predicts a category under which the output will fall. The prediction in such cases is usually a binary answer that's based on "yes" or "no". For instance, it helps with answers like, "will it rain today or not?" "Is this a fruit?" "is this mail spam or not?" and so on. This is attained by referencing a group of data that will indicate whether a certain email falls under the category of spam or not based on specific keywords. This process is known as classification.

Predict a quantity

This system is usually used to predict a value like predicting the rainfall according to different attributes of the weather like the temperature, percentage of humidity, air pressure, and so on. This sort of prediction is referred to as regression. The regression algorithm has various subdivisions like linear regression, multiple regression, etc.

Anomaly Detector Systems

The purpose of a model in anomaly detection is to detect any outliers in the given set of data. These applications are used in banking and e-commerce systems wherein the system is built to flag any unusual transactions. All this helps detect fraudulent transactions.

Clustering Systems

These forms of systems are still in the initial stages but their applications are numerous and can drastically change the way business is conducted. In this system, the user is classified into different clusters according to various behavioral factors like their age group, the region they live in or even the kind of programs they like to view. According to this clustering, the business can now suggest different programs or shows a user might be interested in watching according to the cluster that the said user belongs to during classification.

Main Categories of Machine Learning Systems

In the case of traditional machines, the programmer will give the machine a set of instructions and the input parameters, which the machine will use to compute make some calculations, and derive an output using specific commands. In the case of Machine Learning systems, however, the system is never restricted by any command that the engineer provides, the machine will choose the algorithm that can be used to process the data set and decide the output with high accuracy. It does this, by using the training data set which consists of historical data and output.

Therefore, in the classical world, we will tell the machine to process data based on a set of instructions, while in the Machine Learning world we will never instruct a system. The computer will have to interact with the data set, develop an algorithm using the historical data set, make decisions like a human being would, analyze the information, and then provide an output. The machine, unlike a human being, can process large data sets in short periods and provide results with high accuracy.

There are different types of Machine Learning algorithms, and they are classified based on the purpose of that algorithm. There are three categories in Machine Learning systems:

- Supervised Learning

- Unsupervised Learning

- Reinforcement Learning

Supervised Learning

In this model, the engineers feed the machine with labeled data. In other words, the engineer will determine what the output of the system or specific data sets should be. This type of algorithm is also called a predictive algorithm.

For example, consider the following table:

Currency (label)	Weight (Feature)
1 USD	10 gm
1 EUR	5 gm
1 INR	3 gm

1 RU	7 gm

In the above table, each currency is given an attribute of weight. Here, the currency is the label, and the weight is the attribute or feature.

The supervised Machine Learning system with first we fed with this training data set, and when it comes across any input of 3 grams, it will predict that the coin is a 1 INR coin. The same can be said for a 10-gram coin.

Classification and regression algorithms are a type of supervised Machine Learning algorithms. Regression algorithms are used to predict match scores or house prices, while classification algorithms identify which category the data should belong to.

We will discuss some of these algorithms in detail in the later parts of the book, where you will also learn how to build or implement these algorithms using Python.

Unsupervised Learning

In this type of model, the system is more sophisticated in the sense that it will learn to identify patterns in unlabeled data and produce an output. This is a kind of algorithm that is used to draw any meaningful inference from large data sets. This model is also called the descriptive model since it uses data and summarizes that data to generate a description of the data sets. This model is often used in data mining applications that involve large volumes of unstructured input data.

For instance, if a system is Python input of name, runs, and wickets, the system will visualize that data on a graph and identify the clusters. There will be two clusters generated - one cluster is for the batsman while the other is for the bowlers. When any new input is fed, the person will certainly fall into one of these clusters, which will help the machine predict whether the player is a batsman or a bowler.

Name	Runs	Wickets
Rachel	100	3
John	10	50
Paul	60	10
Sam	250	6

Alex	90	60

Sample data set for a match. Based on this the cluster model can group the players into batsmen or bowlers.

Some common algorithms which fall under unsupervised Machine Learning include density estimation, clustering, data reduction, and compressing.

The clustering algorithm summarizes the data and presents it differently. This is a technique used in data mining applications. Density estimation is used when the objective is to visualize any large data set and create a meaningful summary. This will bring us the concept of data reduction and dimensionality. These concepts explain that the analysis or output should always deliver the summary of the data set without the loss of any valuable information. In simple words, these concepts say that the complexity of data can be reduced if the derived output is useful.

Reinforcement learning

This type of learning is similar to how human beings learn, in the sense that the system will learn to behave in a specific environment, and take actions based on that environment. For example, human beings do not touch fire because they know it will hurt and they have been told that will hurt. Sometimes, out of curiosity, we may put a finger into the fire, and learn that it will burn. This means that we will be careful with fire in the future.

The table below will summarize and give an overview of the differences between supervised and unsupervised Machine Learning. This will also list the popular algorithms that are used in each of these models.

Supervised Learning vs Unsupervised Learning in Summary

Supervised Learning	Unsupervised Learning
Works with labeled data	Works with unlabeled data
Takes Direct feedback	No feedback loop
Predicts output based on input data. Therefore also called "Predictive Algorithm"	Finds the hidden structure/pattern from input data. Sometimes called as "Descriptive Model"
Some common classes of supervised algorithms include: - Logistic Regression - Linear Regression (Numeric prediction) - Polynomial Regression - Regression trees (Numeric prediction) - Gradient Descent - Random Forest - Decision Trees (classification) - K-Nearest Algorithm (classification) - Naive Bayes - Support Vector Machines	Some common classes of unsupervised algorithms include: - Clustering, Compressing, density estimation & data reduction - K-means Clustering (Clustering) - Association Rules (Pattern Detection) - Singular Value Decomposition - Fuzzy Means - Partial Least Squares - Hierarchical Clustering - Principal Component Analysis

Table: Supervised vs. Unsupervised Learning

Let us now look at some examples of where Machine Learning is applied. It is always a good idea to identify which type of Machine Learning model you must use with examples. The following points are explained in the next section:

- Facebook face recognition algorithm
- Netflix or YouTube recommending programs based on past viewership history
- Analyzing large volumes of bank transactions to guess if they are valid or fraudulent transactions.
- Uber's surge pricing algorithm

Steps in building a Machine Learning System

Regardless of the model of Machine Learning, here are the common steps that are involved in the process of designing a Machine Learning system.

Define Objective

As with any other task, the first step is to define the purpose you wish to accomplish with your system. The kind of data you use, the algorithm, and other factors will primarily depend on the objective or the kind of prediction you want the system to produce.

Collect Data

This is perhaps the most time-consuming steps of building a system of Machine Learning. You must collect all the relevant data that you will use to train the algorithm.

Prepare Data

This is an important step that is usually overlooked. Overlooking this step can prove to be a costly mistake. The cleaner and more relevant the data you are using is, the more accurate the prediction or the output will be.

Select Algorithm

There are different algorithms that you can choose, like Structured Vector Machine (SVM), k-nearest, Naive-Bayes, Apriori, etc. The algorithm that you use will primarily depend on the objective you wish to attain with the model.

Train Model

Once you have all the data ready, you must feed it into the machine and the algorithm must be trained to predict.

Test Model

Once your model is trained, it is now ready to start reading the input to generate appropriate outputs.

Predict

Multiple iterations will be performed and you can also feed the feedback into the system to improve its predictions over time.

Deploy

Once you test the model and are satisfied with the way it is working, the said model will be sterilized and can be integrated into any application you want. This means that it is ready to be deployed.

All these steps can vary according to the application and the type of algorithm (supervised or unsupervised) you are using. However, these steps are generally involved in all processes of designing a system of Machine Learning. There are various languages and tools that you can use in each of these stages. Here, you will learn about how you can design a system of Machine Learning using Python.

Machine Learning Scenarios

Scenario One
In a picture from a tagged album, Facebook recognizes the photo of the friend.
Explanation: This is an instance of supervised learning. In this case, Facebook is using tagged photographs to recognize the person. The tagged photos will become the labels of the pictures. Whenever a machine is learning from any form of labeled data, it is referred to as supervised learning.

Scenario Two
Suggesting new songs based on someone's past music preferences.
Explanation: This is an instance of supervised learning. The model is training classified or pre-existing labels- in this case, the genre of songs. This is precisely what Netflix, Pandora, and Spotify do – they collect the songs/movies that you like, evaluate the features based on your preferences and then come up with suggestions of songs or movies based on similar features.

Scenario Three
It focuses on analyzing the bank data to flag any suspicious or fraudulent transactions.

Explanation: This is an instance of unsupervised learning. The suspicious transaction cannot be fully defined in this case and therefore, there are no specific labels like fraud or not a fraud. The model will try to identify any outliers by checking for anomalous transactions.

Scenario Four
This comprises a combination of various models.
Explanation: The surge pricing feature of Uber is a combination of different models of Machine Learning like the prediction of peak hours, the traffic in specific areas, the availability of cabs and clustering is used to determine the usage pattern of users in different areas of the city.

Chapter Summary

Machine Learning is done by feeding the machine with relevant training data sets. Ordinary systems, that is, systems without any artificial intelligence, can always provide an output based on the input that is provided to the system. A system with artificial intelligence, however, can learn, predict, and improve the results it provides through training.

We look at a simple example of how children learn to identify objects, or in other words how a child will associate a word with an object. We assume that there is a bowl of apples and oranges on the table. You, as an adult or parent, will introduce the round and red object as an apple, and the other object as an orange. In this example, the words apple and orange are labels, and the shapes and colors are attributes. You can also train a machine using a set of labels and attributes. The machine will learn to identify the object based on the attributes that are provided to it as input.

Chapter 3: Machine Learning Types

Unsupervised Learning

Unsupervised learning models are further classified into association and clustering problems. Traditional data sets in machine learning worked through labels and followed the logic of "x leads to y." In the case of unsupervised learning models, there are no outcomes and the model just analyzes through the inputs. No training is given to the machines and therefore they are programmed to find the unknown structure in unlabeled data on their own. Unsupervised learning is also known as the training of an artificial intelligence algorithm which is neither classified nor labeled. This allows the algorithm to act on the information without guidance.

Artificial intelligence systems are capable of handling unsupervised learning models for which they generally use a retrieval-based approach. They also have the capability to perform more complex processing tasks when compared to the supervised learning systems. Major applications of unsupervised learning models are self-driving cars, chatbots, expert systems, and facial recognition programs.

Clustering and Association

Clustering

Clustering problems occur when we must discover the inherent groupings in data. In the association rule learning problem, we are required to discover rules that describe large proportions of data. Clustering is one of the major types of unsupervised machine learning algorithms as it runs through the given data and finds any natural clusters if they exist. For example, there could be different groups of users that are differentiated across a few criteria. For example, the criteria can be age, gender or height.

Major types of clustering are defined as follows:

> ➢ Probabilistic clustering: This involves clustering data points into clusters on a specific probabilistic scale.

> ➢ K-means clustering: K-means clustering involves the clustering of data points into a number (k) of mutually exclusive

clusters whereas the method is more focused on selecting the right number for K.

> Hierarchical clustering: Hierarchical clustering classifies data points into parent and child clusters.

> Gaussian mixture models: These are the models featuring a mixture of multivariate normal density components.

> Hidden Markov models: Used to observe data and recover the sequence of states in the model.

> Self-organizing maps: Self-organizing maps use neural networks that learn the distribution and topology of the data.

Generative Models

Generative models are a major part of unsupervised learning models. They create new samples from the same distribution of the given training data. Generally, these models are designed to learn and discover the essence of the given data for generating similar data. The image data set is an example of generative models and has the capability to create a set of images like the given set. Generative models are purely based on the characteristics of unsupervised machine learning models.

Data Compression

Data compression is the process of keeping data sets as small and efficient as possible. Unsupervised learning methods support data compression and perform the task through the dimensionality reduction process. The dimensionality reduction approach assumes that the given data is redundant and utilizes the same concepts as information theory. Also known as the number of columns in a data set, dimensionality reduction can be used to represent information in a data set with only a specific size of the actual content.

Singular value decomposition (SVD) and principal component analysis (PCA) are the main approaches for data compression. PCA is used to find linear combinations that communicate variance in data whereas the SVD model factorizes data into the product of three other, smaller matrices.

For generating new samples from the same distribution, generative models are used, and they are a major part of the unsupervised machine learning approach as well. The model also has the power to automatically learn the features of given data and implement them into the produced data.

Association

Association is a rule-based machine learning method that is used to discover relations between databases. In large sets of data items, we can find interesting associations such as dependencies and relationships through association rules. In data sets, items are generally stored in the form of transactions that can be generated or extracted through external processes and relational databases. Unsupervised machine learning delivers full scalability to association rules algorithms which in return help in accumulating the ever-growing size of data. Furthermore, the evaluation and discovery of interesting associations make is easier for companies and industries to handle big data.

What Is An Association Rule?

An association rule is comprised of a consequent and an antecedent as both of them are list items. The item set is the list for all items in the consequent and antecedent. A typical example of an association rule is market-based analysis, explained as follows:

TID	ITEMS
1	Bread, Milk
2	Bread, Biscuits, Drink, Eggs
3	Milk, Biscuits, Drink, Coke
4	Bread, Milk, Biscuits, Drink
5	Bread, Milk, Drink, Coke

Basic definitions:
Support count for frequency of occurrence of an item set:
Here ({Milk, Bread, Biscuit}) =2.

Association rule for implication expression of the form X->Y, where X and Y are any of the two item sets; for example:
{Milk, Biscuits}->{Drinks}
A frequent item set is an item set having support greater than or equal to the minimum threshold of the data. To better understand the strength of association between the item set s, we can make use of the following metrics:

- Support: The support measure gives an idea of how frequently an item set occurs in all transactions. Taking the example of itemset1 = {Milk} and itemset2 = {Drinks}, there will be more transactions containing milk as compared to those containing drinks. This means that itemset1 has higher support than itemset2.

To calculate the total number of transactions in which the item set occurs, we can make our calculation through the following mathematical equation:

$$Support(\{X\} \to \{Y\}) = \frac{Transactions\ containing\ both\ X\ and\ Y}{Total\ number\ of\ transactions}$$

- Confidence: In unsupervised machine learning, a confidence measure is used to define the likeliness of occurrence of consequent. Given that the cart has antecedents already, a confidence measure could be implemented on all transactions of an item set. For example, of all the transactions containing {James}, how many had {Drink} on them? We can determine by common knowledge that {James} -> {Drink} should be a high confidence rule. The equation to calculate confidence mathematically is defined as follows:

$$Confidence(\{X\} \to \{Y\}) = \frac{Transactions\ containing\ both\ X\ and\ Y}{Transactions\ containing\ X}$$

It does not matter what type of item set is present in the antecedent in the case of a frequent consequent because the confidence for an association rule that has a very frequent consequent will be high.

- Lift: In cases where {X} leads to {Y} in the cart, remember that the value of lift will always be greater than 1. The mathematical equation to represent Lift approach is defined as follows:

$$Lift(\{X\} \to \{Y\}) = \frac{(Transactions\ containing\ both\ X\ and\ Y)/(Transactions\ containing\ X)}{Fraction\ of\ transactions\ containing\ Y}$$

How Does An Association Rule Work?

In machine learning and data mining, association rules are used to represent the probability of relationships for data items in large data sets. Actually, they are if-then statements that are implemented in various models to discover sales correlations in transactional data.

At the basic level, association rule mining involves the use of data patterns and co-occurrence. Machine learning models are used to identify frequent if-then associations which are also known as association rules.

Popular machine learning algorithms that are associated with association rules are SETM, AIS, and A Priori. With the help of the AIS algorithm, the model can generate and count item set s as it scans data. This also helps the algorithm in determining which large item set s contained a transaction and how new candidate items are created by entering large item set s along with other items in the transaction data. Moreover, SETM algorithm is yet another way to generate candidate item set s while the model scans a database.

SETM algorithm is accountable to item set s at the end of the scan because new candidate item set s is generated in the same way as the AIS algorithm. Along with this operation, transaction ID of the generated transaction is saved along with the candidate item set for a sequential structure. An a priori algorithm allows candidate item set s to generate only with the support of large item set s from the previous pass. Large item set s of the previous pass can be joined to generate specific item set s of a larger size.

Association rules are a powerful source in data mining because they are used for predicting customer behaviors and analyzing customer analytics, product clustering, store layout, and market-based analysis. Generally, programmers make use of association rules to build programs that are capable of handling machine learning. Association rule mining actually searches for frequent items in the data set and is used to generate interesting associations and correlations between item set s on both relational and transactional databases.

Density Estimation

Density is the relationship between observations and their probability. Some outcomes of a random variable have low probability density while other outcomes have a higher probability density. For sample data, the probability density is considered as probability distribution which is a major part of machine learning models.

In order to determine the probability density function for sample data, we must determine whether the given observation is unlikely or not. Selecting appropriate learning methods that require input data is best for obtaining a specific probability distribution.

The probability density function for a random sample of data is calculated because the probability density is supposed to be approximated through the probability density estimation process. Density estimation uses statistical models to find underlying probability distribution which in return gives rise to the observed variables.

Kernel Density Estimation

Kernel density estimation is a method of estimating the probability density function of a continuous variable. Being a non-parametric approach, kernel density estimation does not assume any underlying distribution for the variable and is exclusively created with datum.

The probability density function is then estimated by adding all of the given kernel functions and dividing the total number to ensure that every possible value of the probability density function is non-negative and is a definite integral of probability density function over its support sets equal to 1.

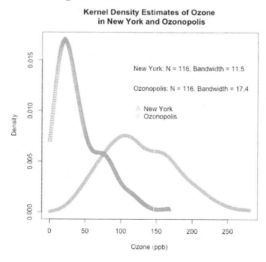

For example:

The above graph is based on the "Ozone" data from the built-in "air quality" data set and the previously simulated Ozone data for Ozonopolis City. It also shows rug plots and how they affect kernel density plots. It can be determined that Ozonopolis has more ozone pollution when compared to New York City, and this is the main reason why the density plots have higher bumps at specific places.

Parameters such as bandwidth and smoothing parameters have the capability to control the scope from data samples which contribute to estimating the probability of a given sample. Moreover, kernel density is also referred to as the Parzen-Rosenblatt window.

Pseudocode for kernel estimation:

- Input: Kernel function K(x), kernel width w, training data instances x1, xn.
- Output: An estimate of the probability density function underlying the training data.
- Process: initialize dens(x) = 0 at all points x

for i = 1 to n:

    for all x:

        dens(x) += (1/n)*(1/w)*K((x-xi)/w)

Latent Variables

What Are Latent Variables?

Latent variables are the variables that are inferred from other variables and are not directly observed. Variables that are observed are also known as directly measured variables and are implemented mathematically to design latent variable models in economics, machine learning, artificial intelligence, psychology, and natural language processing. A latent variable can be observed neither in training nor in the testing phase. These variables can also not be measured on a quantitative scale.

Taking the example of a probabilistic language model of new articles, we can notice that each article x focuses on a specific topic y. For example, sports, economics, politics, or finance articles each focus on their respective topics. By using this knowledge, we can build a more accurate model that can also be expressed as

$p(x \mid t)p(t)$

for which additionally added variable t is unobserved. However, we cannot use learning methods because it is unobserved, and the unobserved variables often make learning a difficult task.

Gaussian Mixture Models

Gaussian Mixture Models are also considered as latent variable models. These models are widely being used in machine learning because they postulate that our observed data is comprised of k clusters. It is assumed that each cluster is specified by $\pi_1,...,\pi_K$; whereas the distribution within each cluster is known as Gaussian.

Chapter 4: How Machine Learning is applied in the Real World

When Does Machine Learning Work?

When faced with a problem that is suitable for the deployment of a computer system, the first thing that you should ask about the problem is whether or not it is rigid and unchanging, or is this something that requires an adaptive system?

At first glance, this seems like a simple question. For example, suppose that you were considering a ballistics program for the military. Ballistics follow the laws of physics. These are precisely known, and so it should be a simple matter to do calculations to get the accurate results that are desired. Hardcoded computer programs can be used to predict how things will work in real situations. Indeed, as soon as Newton's laws were known in the 17th and 18th centuries, armies were using people to do the calculations by hand, and this helped to revolutionize warfare and military efficiency. Today it's done even better and faster using powerful computer technology.

But a recent example shows that there is often more to any situation than meets the eye. Consider an EKG, which can give a doctor a picture of the performance of the heart. An EKG is used to diagnose a heart attack, arrhythmias, and many other problems. It's very narrowly focused, and patterns on the EKG are associated with specific conditions. Health professionals are trained to recognize those patterns, and they can study an EKG chart and determine which patients need medical intervention and which don't. This is as straightforward as the ballistics problem.

However, when artificially intelligent systems were developed using machine learning to study EKGs, it was found that they outperformed doctors by a significant margin. The machine learning systems that have been developed are able to predict which patients will die within a year with 85% accuracy. For comparison, doctors are able to make the same prediction with 65% to at most 80% accuracy. The key here is the difference. When there are EKGs that look completely normal to the human eye—the machine learning system is able to determine that in fact, they are not normal. The engineers that designed the system can't explain it. They don't know why or how the machine learning system makes its predictions. But the way it works, generally speaking, is that the machine learning system is able to detect patterns in the data that human minds cannot detect.

This example serves to illustrate that adaptive learning can be used in nearly every situation. Even in ballistics, there may be many different factors that human engineers have not properly accounted for. Who knows what they are, it could be the humidity, wind, or other factors. The bet is that although line-by-line coded software works very well in deterministic situations, adaptive software that is not programmed and only trained with data will do better.

Complex and Adaptive

When there is any situation where experience—that is exposed to more data—can improve performance, machine learning is definitely called for. If the data is complex, this is another situation where machine learning can shine. Think about how the human mind can handle mathematical problems. Even two-dimensional problems in calculus and differential equations are difficult for most people, and even the smartest people struggle while learning it for the first time. It gets even more difficult when you move to three dimensions, and the more complexity that is added, the harder it is for people to digest.

If you are looking at a data set, you are going to be facing the same situation. If we have a small data set of 20 items, each with 3 fields and an output, a human operator might be able to extract a relationship between the inputs and outputs. They could even do linear regression by hand or plug it into a simple program like Microsoft Excel. Even simply eyeballing the data can reveal the relationships.

But the more data you add to the problem, the less able a human operator is able to determine what the relationships are. The same problem with output but 20 inputs might make it very difficult. If there are no outputs, and you ask human operators to see if there are any patterns in the data, it might become virtually impossible.

One way that we get around complexity in the real world is to program computers in the standard way. This helps humans get around many large data problems and solving problems that would involve tedious work. Consider whether prediction, early efforts at predicting the weather or modeling the climate were based on standard line-by-line coding, using the laws of physics and inputs believed to be important by the operator.

However, when there is a large amount of complexity in a problem, such as predicting the weather, this is a signal that machine learning is probably going to outperform any method by a wide margin. Think about climate modeling. Using conventional techniques, the scientists and programmers are going to make estimates of what factors (such as carbon dioxide level) are important.

But using machine learning, simply training the system on raw data, the system would probably detect patterns in the data that human observers don't even know are there, and it would probably build an even more accurate system that would be better at making future predictions.

To summarize, when you have a problem that is adaptive and complex then it is well suited for machine learning. But there is a third component, and this is big data.

The Role of Big Data

Over the past two or three decades, there has been a quiet revolution in computing power that went unnoticed at first. Developments in technology made it possible to develop more storage capacity, and the costs of this storage capacity have continually dropped. This phenomenon combined with the internet to make it easy for organizations that are large and small to collect enormous amounts of information and store it. The concept of big data was born.

Big data is of course large amounts of data. However, experts characterize big data in four ways.

Simply having a static set of large amounts of data is not useful unless you can quickly access it. Big data is characterized by the "four V's".

- Volume: Huge amounts of data are being created and stored by computer systems throughout the world.

- Velocity: The speed of data movement continues to increase. Speed of data means that computer systems can gather and analyze larger amounts of data more quickly.

- Variety: Big data is also characterized by collection methods from different sources. For example, a consumer profile can include data from a person's behavior while online, but it will also include mobile data from their smartphone, and data from wearable technology like smartwatches.

- Veracity: The truthfulness of the data is important. Do business leaders trust the data they are using? If the data is erroneous, it's not going to be useful.

The key to focus on here is that big data plays a central role in machine learning. In fact, without adequate amounts of accurate (truthful) data that can be accessed quickly, machine learning wouldn't work. The basic fundamentals of machine learning were developed decades ago, but it really wasn't until we moved into the 21st century that the ability to collect and move data around caught up to what was known about machine learning. The arrival of big data is what turned machine learning from an academic curiosity into something real that could be deployed in the real world to get real results.

Where Does Machine Learning Fit In?

Now that we understand the relationship of machine learning to big data, let's see where machine learning fits in with other concepts in computer science. We begin with artificial intelligence. Artificial intelligence is the overarching concept that entails computer systems that can learn and get better with experience. Artificial intelligence can be characterized by the following general characteristics:

- The ability to learn from data.

- The ability to get better from experience.

- The ability to reason.

- It is completely general, as the human brain. So, it can learn anything and can learn multiple tasks.

Machine learning is a subset of artificial intelligence. Rather than being completely general and engaging in human-like reasoning, machine learning is focused on a specific task. There are four major areas of machine learning, and within each of these there are specialties:

- Supervised learning–good for predictions of future outputs.

- Unsupervised learning–good for classifying objects.

- Reinforcement learning–a type of learning that encourages ideal behavior by giving rewards.

- Deep learning–A computer system that attempts to mimic the human brain using a neural network. It can be trained to perform a specific task.

Some Applications of Machine Learning

We have touched on a few ways that machine learning is used in the real world. To get a better feel for machine learning and how it's applied, let's go through some of the most impactful ways that it is being used.

Crimes and Criminal Activity

When you think about machine learning, think patterns. One very practical use of machine learning is exposing a system to past data from criminal activity.

This data can contain many different fields or features. They could include:

- Type of crime committed.

- Location of the crime.

- Time of day.

- Information about the perpetrator or perpetrators.

- Information about the victim.

- Weapons used.

- Day of the week and day of the month.

- Year when the crime occurred.

By studying the data and looking for hidden patterns, a machine learning system can be built to predict the incidents of future crimes. This doesn't mean that it is going to be able to predict a specific crime "there will be a robbery at 615 main street at 6 PM", but rather it will predict overall patterns of criminal activity. This activity might vary in ways that even experienced law enforcement officers are unable to predict—reflect back on the EKG example.

How can this help in the real world? It can help law enforcement agencies deploy resources in a more efficient manner. Even if they don't understand why a given system is making the predictions it's making, they can benefit by moving more law enforcement resources into areas that the system is telling them are going to experience more criminal activity, on the days and at the times when those resources are needed the most. This can help police and emergency personnel respond to crimes more rapidly, and it can also help deter crime with a greater police presence.

Hospital Staffing

Hospital staffing suffers from a similar problem. Human managers attempt to guess when most doctors and nurses are needed and where they should be deployed. While these estimates are reasonably accurate, improvements can be made by deploying a system that uses machine learning. Again, think back to the EKG example—a doctor is pretty good, giving results with 65-80% accuracy. But the machine learning system is even better with 85%, picking out situations the doctors miss. That kind of difference can be a large matter of life or death when it comes to efficiently allocate staff in a large hospital.

To put together systems of this type, large medical organizations tracked the locations and movements of nurses. This allowed them to provide input data to the system, which was able to identify areas of waste. As a simple example, it might have discovered that a large number of nurses were idle on the 7th floor, while there were not enough nurses in a different ward of the hospital, and so patients there were not getting needed attention, and some may have died as a result.

The Lost Customer Problem

For a business, a loyal customer is worth their weight in gold—or far more. A loyal customer is one that is going to return to make repeated purchases. Or even better, they will subscribe. What do you think is more valuable to companies like Verizon and T-Mobile, selling you the phone, or the fact that you sign up for possibly years of regular monthly payments?

Since loyal customers keep business profitable, learning why customers leave is very important. Even just a decade ago, this had to be done using guesswork. But now, vast sums of data have been collected on customers by large corporations. Preventing their customers from switching to a different company is something they are heavily focused on, and machine learning is enabling them to look for patterns in the data that can help them identify why a customer leaves, and even predict when a customer is about to leave.

This data can include basic demographics, usage patterns, and attempts to contact customer support, and so on. The first step where machine learning can be used is that customers that have switched to another company can be identified, and then the system can learn what underlying patterns there are that would enable it to predict what customers are going to leave in the future.

Another way that this data can be used is to study retention efforts. Once a customer is identified that is likely to leave, perhaps they can be offered a special deal. For example, a cell phone company could offer a large discount on a new phone, if they sign up for another two year contract. Or they could offer them free minutes or unlimited data.

Over time, more data is going to be collected. Machine learning can be applied again, this time to determine which methods work the best and what patterns exist—in other words, what method works best for what customers. Maybe you will find that customers of different ages, genders, or living in different locations, or working at different types of jobs, will respond in different ways to inducements offered to retain the customer.

This type of analysis will allow the company to tailor it's responses to each additional customer, improving the odds that they can keep the customer. The customers themselves will be happier, feeling that the company is responsible for their personal needs. The data will also help the company anticipate future market changes and help them adapt using predictive analytics.

Robotics

Using machine learning to develop better and more capable robotics is a huge area of inquiry. Robotics started out simple, performing rote tasks that don't require a huge amount of thought. For example, in the 1980s robotics were introduced on assembly lines to put in a screw or do some other task. The problem then was simple, the robot would perform a small number of exact, rote tasks that could be pre-programmed.

Now robotics are becoming more sophisticated, using cognitive systems to perform many tasks that are tedious but were once thought to be something that only human beings could do. For example, recently robots have been developed that can work as fast-food cooks. This is going to have major implications for unskilled labor because there are two factors at play in the current environment. Activism is pushing up wages for hourly employees doing unskilled labor, while the costs of robotics that can perform the same tasks are dropping. Moreover, the abilities of the robots to perform these tasks continually improve.

A breakeven point is going to be reached. That is the cost of buying and operating a robot will be less than the costs of hiring a human employee. The robot will never waver in efficiency, it won't require the payment of employment taxes, and it's never going to file a lawsuit or allege discrimination. From the employer's perspective, automation is going to be preferable and this trend probably can't be stopped.

Over the past year, many sophisticated robots have been revealed to the public. Boston Dynamics, for example, has built robots that can work in warehouses. They are able to identify packages that need to be moved, pick them up, and then place them where they need to be. At the present time, the only thing preventing widespread adaptation of this type of technology is cost.

A working robot like this has to be able to interact with the environment that it is in, in addition to performing the required task. This means that a sophisticated computer system has to be in place in the robot that includes many machine learning systems. The machine learning systems will include movements required to perform the task, and the ability to avoid running into someone or another robot. Another form of artificial intelligence, computer vision, plays a significant role in the development of robotics, helping it to identify objects (and people) that are in the robot's environment.

Since modern robotics is using machine learning, the ability of the robots to do their jobs and navigate the environments they are in will improve with time.

Chapter 5: Data Mining Techniques

Data mining can be defined as a process for discovering patterns, associations, changes, and events in the data. The patterns discovered must make sense and lead to some advantages, usually leading to those advantages in economics. Useful patterns should allow us to predict new data easily. Data mining is very widespread in the information industry, mainly because of the large amount of data and the immediate need to provide this data. Obtained samples can be used for applications that allow us to analyze, detect fraud, control production, and conduct research ourselves.

Data mining and machine learning are two distinct but related fields. Machine learning provides the technical basis for data mining; it presents methods by which we extract patterns from data in specific collections. So, with machine learning, we teach computers how to handle data and predict what will happen to data. With data mining, however, we gain new knowledge, not just forecasting.

We know many different techniques of data mining. Below, we briefly describe the techniques used in data mining tasks.

Decision Trees

The very beginnings of decision trees date back to the 1960s. The real flourishing of the application, however, reached the decision trees a little later in 1986 when Quinlan introduced the ID3 algorithm, which of course he further upgraded, refined, and renamed to C4.5. The latest version is called C5.0 / See5, which is two programs that can be used to build decision trees. It is designed for two operating systems, UNIX and Windows. Quinlan, because of its great work, is considered to be the first to use a decision tree.

Decision trees are a very popular method, both for classification and for predicting events. They are symbolic methods of machine learning and allow for quick understanding and learning. Their structure is tree-shaped. It is used to subdivide a large number of records into smaller sets (classes) that are sequenced in a certain order.

Decision trees can be thought of as a series of successive questions that aim to classify a record. Each question is followed by a new question, which depends on the previous answer. This process lasts until we determine which group the particular record belongs to. A series of these questions and answers can be graphically depicted as an inverted decision tree.

It is composed of:

- Root node - does not have any input signal and does not have any output signal or has multiple output signals;
- Internal node - has one input signal and two or more output signals;
- Terminal node - has exactly one input signal and no output.

The record enters the decision tree at the source node. A test is made here to determine which inner node the record will enter. Different algorithms are used for this test, but the principle is always the same: choose the test that best discriminates between the target classes. This process is repeated until the record reaches the final node. Each path from the source to the final node is unique and represents a rule that is then used for classification. All records ending in the same end node belong to the same class.

The decision tree consists of the root or node at the top of the tree and the individual, internal nodes, branches, and leaves. The inner nodes are labeled with attribute names and contain some condition that divides the learning set into smaller sets. The branches or links that are marked with possible values for the attribute are a subset of the attribute values. The leaves are outer nodes that correspond to the classes. The path from the root to a leaf corresponds to one decisive rule. The choice of attributes depends on the multitude of learning patterns, the circumstances, the ability to perform measurements. A decision is an event that will happen in the near future if we decide on it. With decision trees, we can try to predict an event that will occur when a decision is made, or with it, we can find the other most optimal options to reach our goals. With them, we are creating ever easier rules that work in as many patterns as possible, and we must remember that we do not become too general.

The decision tree can be built with the help of a learning crowd consisting of some patterns. The properties of the samples are described by a multitude of attributes or with the attributes and classes they belong to (outcome). In this case, one teaching sample is considered to belong to one class. When constructing, we must also keep in mind that multiple patterns that are described with the same attribute vector may not have different choices. If the last two conditions are met, then we are talking about consistent learning patterns. We also know inconsistent learning patterns that are a sign that we made a mistake in the attribute measurement phase. We also have two different attributes in the learning set. They can be discrete with which to build a tree directly, or we have continuous attributes that must be mapped to a discrete form before construction. The process by which decision trees are built, in other words, is called induction.

There are several different types of decision trees, among which we will mention the most famous binary decision tree and the regression decision tree. We recognize a binary decision tree, where each internal node has exactly two successors. Any "ordinary" decision tree can be converted to a binary. They provide us with better classification accuracy.

Regression trees use both continuous and discrete attributes, the number of which must be known in advance. The algorithm for constructing regression trees is very similar to the algorithm for constructing ranking decision trees. Such trees can be used to determine the value of the dependent instance variable of new cases. Alternatively, we can build a binary regression tree that allows us to evaluate the attributes for the variance difference. This produces smaller trees, which in turn give us better accuracy.

The decision tree could be briefly said to be like a flowchart with flow blocks. One disadvantage is that decision trees are very difficult to edit when the data volume increases and the number of branches is too large.

We know of a large number of different tools that are specialized in the construction of decision trees, such as Weka, See5, Gattree, which we will describe later. There are, of course, many other tools.

Bayes Classifier

The Bayes classifier is one of the statistical classifiers. These work by predicting the likelihood that an individual record belongs to a particular class. The Bayes classifier uses the Bayes theorem as its basis. The Bayesian theorem is named after its inventor, Thomas Bayes. Bayes was a British mathematician and Presbyterian monk who lived in London in the 18th century. He wrote his findings on probability in the paper "Essay towards Solving a Problem in the Doctrine of Chances," published two years after his death (International Society for Bayesian Analysis, 2009). Two methods are used in practice:

- The naive Bayesian classification assumes that the influence of attributes on a record class is independent of other attribute values of the same record. This assumption is made to simplify the amount of conversion required. Because of this assumption, we also say that this classifier is naive.

- Bayesian belief networks are graphical models that, in contrast to the previous method, also take into account the dependencies between the variables.

Classification Rules

Rule-based classification is a technique based on the rules of type IF-THEN. IF the part expresses the conditions that must be met (e.g., age = 29, number of children = 4), then the part represents the "consequences" of this rule (e.g., married = yes). An example of this rule if - then, would be:

IF age = 29 AND number of children = 4 THEN married = yes

Commonly, there are two ways of obtaining classification rules:

- A rule derived from decision trees works by first summarizing rules from decision trees and then removing rules that do not improve the accuracy of the rule. This is done using the C4.5 algorithm.
- Rules obtained by sequential covering algorithm. The algorithm works by creating a rule (in the direction general -> specific) for the records that match it. Then these rows are removed from the data set, and the process is repeated until the records run out. The most popular algorithms are CN2, AQ, and RIPPER.

Random Forests

Random forest is a versatile machine learning method that can perform prediction and classification tasks. It also addresses missing values, extreme values, the importance of variables, and other essential data mining steps. The forest chooses the classification that has the most votes (more than the other trees in the forest) and, in the case of regression, uses the average of the results of the different trees. The Random Forest method can work with thousands of input variables, identifying the most important ones. Because of this, it falls under dimension reduction methods (dimension reduction means eliminating irrelevant input variables). Cases of use:

• Customer switching to competing mobile operators (people with certain characteristics and usage patterns).

• Prediction of stock movements (forecasting by market, past growth, the value of other shares, etc.).

Positive features of use are the effective performance on large amounts of data and the good estimation of missing values. Negative features: Unlike the decision trees, the results are difficult to interpret. Also, the frequency of cases of overlap.

Support Vector Method

The first SVM research was published in 1992, and its design dates back to 1960. SVM is a new method that has received a lot of attention lately. The expert public is enthusiastic about this method mainly because of its ability to form complex nonlinear and multidimensional planes that delineate groups of objects perfectly. Another advantage is that they give a concise description of the model obtained. A major disadvantage of the SVM method is that it takes a very long time to train them and require a lot of processing power.

SVM is a controlled learning algorithm that is used for both classification and prediction. SVM works by creating a hyperplane or multiple hyperplanes in a multidimensional space, delimiting individual groups. The plane delimiting the two groups delimits them so that there is a maximum distance (W) between them. In this way, overfitting is minimized. The concept is shown in the figure below in two-dimensional space.

The SVM model presents the data as points in space, arranged so that the cases of the separated categories are divided by a clear distance as wide as possible. The method performs well in prediction, redundancy removal, and classification. Examples of use:

- Face recognition (on which part of the image is the face).
- Classification of images.
- Bioinformatics (protein classification).
- Handwriting recognition.

Positive features of use: are the efficiency in multiple spaces and memory efficiency. Negative ones are Poor performance on large amounts of data and overlapping target classes.

Neural Networks

Neural networks were conceptually designed in 1943 by neuropsychologist Warren McCulloch and logician Walter Pitts to try to explain how a neuron works in the human body. Although their purpose was to explain the functioning of the human brain, this model has been shown to offer a new approach to the problem-solving outside of neurobiology. In the 1950s, perceptrons were developed, the design of which was based on the work of McCulloch and Pitts. They had some success in the lab with them but generally did not live up to their expectations. The reasons for this were the very low capacity of the computers of the time, as well as the theoretical shortcomings of these simple neural networks. These deficiencies were eliminated in 1982 by John Hopfield with a new way of learning neural networks. He called it "backpropagation," which could be explained as reverse or backward learning. "Backward learning" is a way of updating weights at higher levels before updating weights at lower levels. This procedure allows us to use a higher-level neuron error to evaluate a lower-level neuron error.

The discovery of "learning backward" triggered a renaissance in neural network exploration. In the 1980s, their use moved from laboratories to commercial areas, which were used almost everywhere. Some examples of use include real-time detection of credit card fraud, identifying numbers written on checks, evaluating real estate values, detecting patients' illnesses, and the like.

The advantages of neural networks are:

•   We can handle a wide range of problems. They are very flexible.

•   They also perform well in complex areas.

•   Can use continuous and discrete variables.

•   They are very widespread and easily accessible. They are found in almost all commercial data mining applications.

Neural networks are very flexible. They can be used to predict continuous variables - in this case, they perform prediction, but we can use them to predict discrete variables - in this case, they perform classification. If we rearrange the neurons in the network a little, neural networks can be used to detect groups in the data.

Neural networks do well with continuous and discrete variables as input or output. A lot of work, however, has to be invested in data preparation. Because neural networks only work with numbers from 0 to 1, all variables must be reduced to this form.

The disadvantages of neural networks are:

•   Input data must be in the form of a number in the range 0 to 1.

•   We cannot explain why the results are as they are.

•   There is a possibility that they may not find the optimal solution.

For their operation, neural networks require input data in the form of a number in the range from 0 to 1. This, of course, requires a lot of work required to transform the data. Choosing a technique for data transformation is also crucial and can have a significant impact on mining results.

The operation of neural networks can be imagined as a "black chest" in which we input and get a result. However, we cannot see how we came to this result, nor can we pull out the reins as the Prime Minister can in associative rules or decision trees. Figure 6 illustrates a neural network with input, hidden, and output layers.

We can see that we know the input layer that we define ourselves, we know the output layer, which is the goal of the whole operation, but we do not know the hidden layer, and therefore, no rule can be extracted from it.

Neural networks perform well for most predictive and classification tasks, where the results are more important than understanding how the model works. Neural networks also work well in clustering. We can use the SOM (Self-Organizing Map) technique for this. While using this technique, we get individual groups, but we do not know why the elements within groups are similar, and we need to analyze them using the average values of the elements in each group.

The only case where neural networks do not perform well is when we have elements with hundreds or thousands of attributes. Due to a large number of neural network attributes, it is difficult to find the pattern, and the time required to train the model is markedly increased. One way to solve this problem is to combine neural networks using decision trees that behave in identifying important attributes.

Chapter 6: Setting up the Python environment for Machine Learning

Installation Instructions for Python

WINDOWS

1. From the official Python website, click on the "Downloads" icon and select your operating system.

2. Click on the "Download Python 3.8.0" button to view all the downloadable files.

3. On subsequent screen, select the Python version you would like to download. We will be using the Python 3 version under "Stable Releases." So, scroll down the page and click on the "Download Windows x86-64 executable installer" link as shown in the picture below.

- Python 3.8.0 · Oct. 14, 2019

  Note that Python 3.8.0 *cannot* be used on Windows XP or earlier.

  - Download Windows help file
  - Download Windows x86-64 embeddable zip file
  - Download Windows x86-64 executable installer
  - Download Windows x86-64 web-based installer
  - Download Windows x86 embeddable zip file
  - Download Windows x86 executable installer
  - Download Windows x86 web-based installer

4. A pop-up window titled "python-3.8.0-amd64.exe" will be shown.

5. Click on the "Save File" button to start downloading the file.

6. Once the download has completed, double click the saved file icon and a "Python 3.8.0 (64-bit) Setup" pop window will be shown.

7. Make sure that you select the "Install Launcher for all users (recommended)" and the "Add Python 3.8 to PATH" checkboxes. Note – If you already have an older version of Python installed on your system, the "Upgrade Now" button will appear instead of the "Install Now" button and neither of the checkboxes will be shown.

8. Click on "Install Now" and a "User Account Control" pop up window will be shown.

9. A notification stating, "Do you want to allow this app to make changes to your device" will be shown, click on Yes.

10. A new pop up window titled "Python 3.8.0 (64-bit) Setup" will be shown containing a set up progress bar.

11. Once the installation has been completed, a "Set was successful" message will be shown. Click on Close.

12. To verify the installation, navigate to the directory where you installed Python and double click on the python.exe file.

MACINTOSH

1. From the official Python website, click on the "Downloads" icon and select Mac.

2. Click on the "Download Python 3.8.0" button to view all the downloadable files.

3. On subsequent screen, select the Python version you would like to download. We will be using the Python 3

- Python 3.7.5 - Oct. 15, 2019
  - Download macOS 64-bit/32-bit installer
  - Download macOS 64-bit installer
- Python 3.8.0 - Oct. 14, 2019
  - Download macOS 64-bit installer
- Python 3.7.4 - July 8, 2019
  - Download macOS 64-bit/32-bit installer
  - Download macOS 64-bit installer
- Python 3.6.9 - July 2, 2019

version under "Stable Releases." So, scroll down the page and click on the "Download macOS 64-bit installer" link under Python 3.8.0, as shown in the picture below.

4. A pop-up window titled "python-3.8.0-macosx10.9.pkg" will be shown.

5. Click "Save File" to start downloading the file.

6. Once the download has completed, double click the saved file icon and an "Install Python" pop window will be shown.

7. Click "Continue" to proceed and the terms and conditions pop up window will appear.

8. Click Agree and then click "Install."

9. A notification requesting administrator permission and password will be shown. Enter your system password to start installation.

10. Once the installation has finished, an "Installation was successful" message will appear. Click on the Close button and you are all set.

11. To verify the installation, navigate to the directory where you installed Python and double click on the python launcher icon that will take you to the Python Terminal.

LINUX

For Red Hat, CentOS or Fedora, install the python3 and python3-devel packages.

1. From the official Python website, click on the "Downloads" icon and select Linux/UNIX.

2. Click on the "Download Python 3.8.0" button to view all the downloadable files.

3. On subsequent screen, select the Python version you would like to download. We will be using the Python 3 version under "Stable Releases." So, scroll down the page and click on the "Download Gzipped source tarball" link under Python 3.8.0, as shown in the picture below.

- Download Gzipped source tarball

- Download XZ compressed source tarball

- Python 3.8.0 - Oct. 14, 2019

  - Download Gzipped source tarball

  - Download XZ compressed source tarball

- Python 3.7.4 - July 8, 2019

  - Download Gzipped source tarball

  - Download XZ compressed source tarball

4. A pop-up window titled "python-3.7.5.tgz" will be shown.

5. Click "Save File" to begin downloading the file.

6. Once the download has finished, double click the saved file icon and an "Install Python" pop window will appear.

7.   Follow the prompts on the screen to complete the installation process.

Getting Started

With the Python terminal installed on your computer, you can now start writing and executing the Python code. All Python codes are written in a text editor as (.py) files and executed on the Python interpreter command line as shown in the code below, where "nineplanets.py" is the name of the Python file:

"C: \Users\Your Name\python_nineplanets.py"

You will be able to test a small code without writing it in a file and simply executing it as a command line itself by typing the code below on the Mac, Windows or Linux command line, as shown below:

"C: \Users\Your Name\python"

In case the command above does not work, use the code below instead:

"C: \Users\Your Name\py"

Indentation – The importance of indentation, which is the number of spaces preceding the code, is fundamental to the Python coding structure. In most programming languages indentation is added to enhance the readability of the code. However, in Python the indentation is used to indicate execution of a subset of the code, as shown in the code below

If 7 > 2:

    print ('Seven is greater than two')

Indentation precedes the second line of code with the print command. If the indentation is skipped and the code was written as below, an error will be triggered:

If 7 > 2:

print ('Seven is greater than two')

The number of spaces can be modified but is required to have at least one space. For example, you can execute the code below with

427

higher indentation but for a specific set of code same number of spaces must be used or you will receive an error.

If 7 > 2:

..print ('Seven is greater than two')

Adding Comments – In Python comments can be added to the code by starting the code comment lines with a "#," as shown in the example below:

> #Any relevant comments will be added here

> print ('Nine planets')

Comments serve as a description of the code and will not be executed by the Python terminal. Make sure to remember that any comments at the end of the code line will lead to the entire code line being skipped by the Python terminal as shown in the code below. Comments can be very useful in case you need to stop the execution when you are testing the code.

> print ('Nine Planets')#Comments added here

Multiple lines of comments can be added by starting each code line with "#," as shown below:

> #Comments added here

> #Supplementing the comments here

> #Further adding the comments here

> print ('Nine Planets')

Python Variables

In Python, variables are primarily utilized to save data values without executing a command for it. A variable can be created by simply assigning desired value to it, as shown in the example below:

```
> A = 999
> B = 'Patricia'
> print (A)
> print (B)
```

A variable could be declared without a specific data type. The data type of a variable can also be modified after its initial declaration, as shown in the example below:

```
> A = 999 # A has data type set as int
> A = 'Patricia' # A now has data type str
> print (A)
```

Some of the rules applied to the Python variable names are as follows:

1. Variable names could be as short as single alphabets or more descriptive words like height, weight, and more.

2. Variable names could only be started with an underscore character or a letter.

3. Variable names must not start with numbers.

4. Variable names can contain underscores or alphanumeric characters. No other special characters are allowed.

5. Variable names are case sensitive. For example, 'weight,' 'Weight' and 'WEIGHT' will be accounted as three separate variables.

We recommend reading our first book of the series to learn the basics of Python.

Chapter Summary

This provides the core instructions to be followed in downloading and installing Python on your operating system. The latest version of Python released in the middle of the 2019 is Python 3.8.0. Make sure to download and install the most recent and stable version of Python at the time.

Chapter 7: Top libraries and modules for Python Machine Learning Applications

NumPy

One of the most fundamental open source libraries for Python when it comes to computation is NumPy (Numerical Python).
It offers primarily mathematical functions that perform operations on data. It has the characteristic of very fast and efficient execution of functions.

Pandas

Pandas is a free open source library created by Wes McKinney that works with tagged and relational data.
It is designed for fast and easy data manipulation, aggregation, filtering, etc. It has two data structures, Series and DataFrame.

Matplotlib

The Matplotlib Library has been designed for easy and powerful data visualization and has been around for over 16 years. The latest stable version is 3.1.1, which was released in July 2019. This library, along with NumPy and Pandas, is a very serious competitor to scientific, well-known tools (but quite expensive) such as MatLab and Mathematica.
What is a disadvantage of Matplotlib is that it is a quite low level which means that it takes more lines of code to come up with some more advanced visualizations which means that it takes a lot more effort and time than when using high level paid tools, but certainly the effort is worth the try.

Seaborn

The Seaborn library is primarily focused on creating attractive visualizations of statistical graphs such as heat map views, data distributions, etc. This library is based and dependent on the Matplotlib library.

Bokeh

Another library focused on visualizations. Unlike Seaborn, Bokeh is independent of Matplotlib.
The main focus of this library is the interactivity and presentation of visualizations through modern web browsers like D3.js.

Altair

Altair is a newer library for statistical visualization. With minimal code, highly effective and beautiful visualizations can be produced, and the API of this library is simple and built on the powerful Vega-Lite.
New chart types and visualizations can be expected in the future as it is under constant and active development.

SciKit-learn

SciKit-Learn is an open-source library that is currently one of the industry standards when it comes to machine learning in Python. It can also serve as a very effective tool for data mining and data analysis. It was built based on NumPy, SciPy, and Matplotlib.
The library combines quality code and good documentation, ease of use, and excellent performance. Has classification, regression, clustering, preprocessing, etc. It is designed to work with Numpy and SciPy libraries. It has machine learning algorithms and can integrate machine learning with production systems very easily.

MlPy

MlPy is a Python machine learning library built on Numpy-SciPy libraries. It offers a wide range of machine learning methods, both supervised and unsupervised. This is a multi-platform, which means it works on both Python v2 and Python v3.

Scrapy

Scrapy is a library for making crawling programs known as spider bots for delivering structured data such as contact information or URLs from the internet.

It is open-source, written in Python, and originally designed strictly for scraping. Over time, Scrapy has evolved into a mature framework with the ability to collect data from the API and also act as a general crawler.

Statsmodels

Statsmodels is a Python library that allows users to research data, evaluate statistical models, and perform statistical tests. It has a comprehensive list of descriptive statistics, statistical tests, and is available for different types of data.

It provides extensive visualization features and is designed primarily for statistical analysis and is characterized by high performance, especially when used with large data sets.

NLTK

The abbreviation for this library means The Natural Language Toolkit and is used for the frequent tasks of natural language statistical processing in Python. It is designed primarily to facilitate the learning and exploration of NLP and its related fields (linguistics, cognitive science, and artificial intelligence).

Gensim

Gensim is an open-source library for Python that implements tools that work with vector modeling. It achieves efficiency by using the NumPy data structure and SciPy operations, and it works great when it comes to big texts.

Gensim is used for raw and unstructured digital texts and implements various unsupervised (unsupervised) algorithms.

NumPy (Numerical Python)

Numpy is a library made up of multidimensional vectors and the functionality needed to work with them. It is most commonly used for mathematical and logical operations on vectors, Fourier transforms, form manipulations, operations required for linear algebra, and also contains a random number generator. We will install the package with the following command:

$ pip install numpy

The most important component is ndarray (multidimensional vector), which is a 'container' containing elements of the same size and type. Each element is a data type, dtype object. Calling the shape parameter checks the dimensions of the vector while calling the dtype parameter reveals what type of data is inside the vector.

A data type object is a fixed block of memory intended for a vector. The size of the block it occupies depends on the data type, data size, byte order and the like. NumPy contains numerous mathematical operations, such as trigonometric (sin, cos, arcsin, arccos, degrees) arithmetic (add, subtract, multiply, divide, mod), operations with complex numbers (real, imag, conj, angle) and many others. It also contains statistical operations such as minimum, maximum, percentiles, arithmetic mean, standard deviation and variance of given elements in a vector. It also makes it easier for us to deal with linear structures, such as matrices. It contains operations for matrix multiplication, determinant computation, inverse search, and the like.

The combination of NumPy and Matplotlib gives us the ability to visualize functions and graphs. Using numpy, we define the x and y axes while drawing a graph with the pyplot () function, which we imported from matplotlib. We can further 'decorate' the graph by adding a third parameter where the first part is the look of the function, and the second is the color of the function, all in quotes.

The Pyplot function has the ability to draw bar graphs (bar). The most common visualization of frequency and data distribution is a histogram, which we can also display using numPy, defining values, and intervals.

Matplotlib

The Matplotlib library is used to visualize data and draw 2D graphs. Displaying multiple datasets will be obtained by adding another pair of parameters to the plot () method for another dataset, while achieving multiple sub-graphs on the same 'figure' is achieved by the subplot method (number of rows, number of columns, ordinal number)).

For drawing multiple graphs, this method is not optimal, and therefore the subplots (nrows, ncolumns, figsize) method is used. We approach each graph as in a matrix system (a [row: column)).

The bar chart contains two other important methods: legend () - shows what color is displayed and xticks () - adjusts the column labels on the x-axis. We use the pie () method for a pie chart. We can shade the diagram and highlight some elements with the explode parameter.

To visualize the scatter graph (regression problems), we use the scatter (height, weight), xlim (int), and ylim (int) methods. We can also visualize the scatter graph over a 3D model, including the mplot3d module, and assign the projection = '3d' axes parameter to the method.

Scikit-Learn

Scikit-Learn, the library most used with NumPy and Pandas, is intended for data modeling. Used for classification, regression, and grouping. Scikit comes with several standard datasets, such as iris and digits.

Data is always in a 2D matrix (n samples, n features), regardless of the original format. We can also visualize the data using the subplot method.

Linear regression predicts the value of the dependent variable y versus the independent variable x. We load the pandas' data with the read_csv (path) function, we check the shape of the dataset with the shape attribute, which returns us the number of data and features, and we print the actual data using the head () method. We can see the statistics of the set by calling the describe () method.

The first step to linear regression is to separate the data into dependent and independent variables and to separate the learning data and the test data. We then create a linear regression object, to which we subsequently describe the learning model. We perform testing by calling the predict () method.

All we have to do is sketch two linear regression models, one with learning data and the other with test data.

Pandas

Pandas is one of the most used libraries for data analysis because taking data from a CSV, TSV file, or SQL database creates an object called a data frame.

Data Frame

Data is loaded as python components, local files, or via URL. and then converted to a data frame. pd.read_filetype () pd.DataFrame () When working on data, we can also save the data frame with the to_filetype (filename) method. We can also create our own data frame by storing key-value pairs in the dictionary. in this case, the name of the column-value column, and then assign that dictionary to the data frame.

Individual data is accessed with the loc attribute 'key.'

Read from Database

To read from the database, we need to establish a connection to the database, so first, we need to download the python library for the database. For example, to use the SQLite database, we need to install the pysqlite3 library. import sqlite3 con = sqlite3.connect ("database.db") df = pd.read_sql_query ("SELECT * FROM table_name", con)

Most important Data Frame Pperations

When loading a dataset, it is useful to see the first few rows, to see the structure of the data. This can be achieved by the head (n) method, which by default, throws out the first few lines, but we can also assign an arbitrary number to it. To see the last few lines, we use the tail (n) method. Calling the info () method, we get basic information about the dataset, such as the number of variables and features, the type of data, and how much space it takes up.

The shape attribute gives us the number of rows and columns, which is useful for sorting and filtering data (to find out how much data we have left after). The append () method returns a copy of the original data frame with new values, and the drop_duplicates () method returns a copy with the duplicates removed.

If we do not want to constantly assign a modified data set to a new data frame, then we use the inplace = True argument within the methods we make changes to. temp_df.drop_duplicates (inplace = True)

Drop_duplicates () also contains a keep argument with three options:

First - delete all duplicates except the first

Last - delete all duplicates except the last

False - delete all duplicates

If we want to rename the column, we will use the dictionary rename () method.

To check how many null values we have in a dataset, we use the isnull () and sum () method aggregation.

data.isnull (). sum ()

To remove null values, we use the dropna () method, which will delete every row that has at least one null value and save a new dataset to a copy. In addition to rows, we can also delete columns with null values, adding the axis = 1 parameter. In order not to delete columns containing valuable data, just because they have one / two null values, we will use a technique called Imputation ('padding'). It is often used by technical engineers to replace each null value with the arithmetic mean or median of that column. We fill in the blank values with the fill (new value) method. In the Scikit-Learn - Basics section, we saw that the describe () method sees the distribution of continuous variables. In addition to the entire dataset, it can also be used on a categorical feature (column), where we then get a detailed view of unique values, categories, and the like. The correlation coefficient shows us the relationship between the two variables. If it is positive, it means that if one variable rises and the other rises (one falls and the other falls), and if negative, it means that their relationship is such that if one falls, the other rises. The number 1.0 indicates the best correlation between the variables. We use the corr () method to print the correlation coefficients.

Indexing and a Conditional Selection

To access the row, we use two arguments: loc and iloc. loc is used to search by name, and iloc to search by index. We can select multiple lines by writing from-to, the same as in Python lists. The only difference is that when indexing with names (loc), the last specified edge is included, which Python lists and iloc do not do.

If we want rows with a specific value, we apply a Boolean condition. It returns TRUE or FALSE, depending on whether or not it satisfies the condition. If we want to filter the data frame to have no FALSE value, then we set the condition to the data frame itself.

For more complicated conditions, we use both (&) and or (|) operators or simplify them with the isin () method.

Applying Functions to Datasets

When we want to change something or add a new column, the values of which are obtained using the original dataset, it is optimal to write a function that we will later apply to the data frame method apply ().
Graphs
In combination with Matplotlib, Pandas can visualize data frames with a variety of graphs. The boxplot is based on a set of different numbers describing the data set (minimum, first quartile Q1, median, third quartile Q3, and maximum). It tells us how symmetrical the data set is, what the distribution density is, and the like. There are two ways to show a rectangular diagram. The first is to put kind = "box" in the plot method or call the boxplot method (column = "column_name" by = "sort_by_column").
Interesting Facts
When we do not have a data file already saved and when we do not want to waste time-saving the file, Pandas offers the ability to read data from the clipboard. So we can copy the table from an Excel file and call the read_clipboard () method and Pandas will load it like any other file. After working with the data, we can immediately save the file in gzip, zip, or bz2 format by adding the compression = "gzip / zip / bz2" parameter to the to_filetype (filename) method. We can use CSS properties to style the tables.

Chapter Summary

Libraries are a set of features and modules, designed to make it easier to write code so that we use existing functionality and do not have to write it ourselves. The most commonly used Python libraries for data science/analysis are NumPy, Pandas, SciPy, Scikit Learn, Matplotlib, Seaborn, and many others. We can include the library by assigning it a nickname, or including all or some of the functionality in the library without aliases after we can only call them a function name.

Conclusion

Python is a widely-used programming language for different applications and in particular for machine learning. We cover the basic Python programming as well as a guide to using Python libraries for machine learning.

We present machine learning applications using real datasets to help you enhance your Python programming skills as well as machine learning basics acquired through the book. These applications provide examples of developing a machine learning model for predictions using linear regression, a classifier using logistic regression and artificial neural network. Through these applications, examples of data exploration and visualization using Python are presented.

Machine learning is an active research subject, in particular, artificial neural networks. Nowadays, machine learning is used in every domain, such as marketing, health care systems, banking systems, stock market, gaming applications, among others. Our objective is to provide a basic understanding of the major branches of machine learning as well as the philosophy behind artificial neural networks. The book also aims at providing Python programming skills for machine learning to beginners with no previous programming skills in Python or any other programming language.

Once you have acquired the skills and understood the reasoning behind machine learning models presented here, you will be able to use these skills to solve complex problems using machine learning. You will also be able to easily acquire other skills and use more advanced machine learning methods.

By now you are able to apply various machine learning algorithms for supervised learning, unsupervised learning and reinforcement learning. This knowledge is beneficial when you want to explore other problems regarding machine learning. But we then, it is important to give you more words of advice especially on how to approach a problem regarding machine learning. Some of these procedures have been outlined in the sections below:

Approaching a machine learning problem

One must be very orderly especially when approaching a machine learning problem for the first time. Endeavor to answer the following questions: in case you are interested in fraud detection, then consider:

- What and how do I measure for a working fraud prediction model?

- Do I have the knowledge to evaluate a specific algorithm?

- What is the final implication (businesswise or academically) of my model in case I will be successful?

Humans in the Loop

Identify and assign roles of each and every human participant in your model appropriately.

Move from prototype to production

Form porotypes built using libraries features of scikit-learn, engage the production team who should be working with programming languages including: Python, R, Go, Scala, C#, C++, and C languages.

Testing production systems

Before you release your products into the market make sure you carry out proper testing to ascertain whether your product is working properly. It is in this phase that you can identify various faults and even points of improvement.

Next step: In this case, we provide a perfect illustration for machine learning that at the end of it you expect to be a powerful machine learning expert. However, if you need more information, some sources have been included below for reference purposes.

1. Feel free to consult form other literature about those concepts that may not be very clear to you. Some good books may include: Tibshirani, and Friedman's book The Elements of Statistical Learning and An Algorithmic Perspective by Stephen Marsland (Chapman and Hall/CRC)

Machine learning packages: Currently there are a number of machine learning packages in the market that are meant to boost your understanding of various algorithms. Some of the most common packages are scikit-learn (most interactive and obviously favorite), statsmodels package, vowpal wabbit (vw) and finally mllib (a Scala library built on Spark)

Perform ranking on recommender systems and other kinds of machine learning systems in order to identify the best of the bests.

Also carry out probabilistic modeling, probabilistic programming and interfacing in order to increase your knowledge on machine learning algorithms.

Other more recommended topics though very complex would be Neural networks, Large Data sets scaling and Deep Learning.

It is my hope that you are now convinced about machine learning, the new niche for building new applications and objects (project manipulation). Keep digging into this new area and never give up, a lot more is yet to come.

The End

Made in the USA
Columbia, SC
20 March 2021